WEATHER

WEATHER

THE ULTIMATE BOOK OF METEOROLOGICAL EVENTS

FOREWORD BY DR. D. JAMES BAKER

FROM THE CONTRIBUTORS OF THE *WEATHER GUIDE CALENDAR*

Weather: The Ultimate Book of Meteorological Events

Photography credits and copyrights page 254.

For information, write Accord Publishing, a division of Andrews McMeel Publishing, LLC,
1404 Larimer Street, Suite 200, Denver, CO 80202

Packaged by Jennifer Barry Design, Fairfax, CA
Design and editing: Jennifer Barry
Layout production: Kristen Hall
Text editor: Blake Hallanan
Photography coordinator: Laura Del Fava
Foreword by D. James Baker
Essays on pp. 27, 57, 93, 133, and 185 by Walter Lyons, CCM

Printed in China

07 08 09 10 11 WKT 10 9 8 7 6 5 4 3 2 1

ISBN-13: 978-0-7407-6989-4
ISBN-10: 0-7407-6989-8

Library of Congress Cataloging-in-Publication Data:

Weather : the ultimate book of meterological events / from the contributors of the Weather Guide Calendar.
p. cm.
Includes bibliographical references and index.
ISBN-13: 978-0-7407-6989-4 (alk. paper)
ISBN-10: 0-7407-6989-8 (alk. paper)
1. Weather—Popular works. 2. Meteorology—Popular works.

QC981.2.W47 2007
551.5—dc22 2007038988

PRECEDING PAGES 2–3: *This isolated rainstorm with cloud-to-ground lightning was taken at sunset in southeast Arizona, an area known for its spectacular sunsets. The red coloring seen in this photograph may have been created by a lack of moisture in the atmosphere, or by dust particles stirred up by the wind as a result of the thunderstorm. It might also have been attributed to light scattered by ash produced by the eruption of Mt. Pinatubo in the Phillipines on June 1, 1991—the most violent volcanic eruption in the 20th century. Photo: Warren Faidley/Weatherstock*

PRECEDING PAGES 4–5: *A storm approaching from the south near Steamboat Springs, Colorado, creates the perfect conditions for a majestic double rainbow. Rainbows are optical and meteorological phenomena that cause a spectrum of light to appear in the sky when the sun shines through droplets of moisture in the atmosphere. They take the form of a multicolored arc, with red on the outer part of the arc and violet on the inner part of the arc. More rarely, a secondary rainbow is seen which is a fainter arc outside the primary arc with colors in the opposite order. Photo: Rod Hanna*

PRECEDING PAGES 6–7: *Lenticular ("lens shaped") clouds, such as this one over the Colorado Rockies, are most frequently seen near mountain ranges. Wind blowing over the mountains sets up a stationary wave pattern on which the cloud forms: these formations are sometimes known as standing wave clouds. In a few rare cases, the lens or saucer shape appears so unusual that these clouds have been reported as flying saucers. Unusual coloration, such as shown here from the setting sun, can make them especially beautiful. Photo: Chuck Conway*

PRECEDING PAGES 8–9: *Valley fog clears after a recent storm to unveil Sleeping Giant, a landmark mountain near Steamboat Springs, Colorado. Photo: Rod Hanna*

LEFT: *Lightning doesn't always require a storm, as this photo of the Japanese volcano Mount Sakurajima, about 50 miles north of Iwojima attests. Eruptions of the Sakurajima (Cherry Island) volcano spewed tons of ash into the air. The lightning resulted from friction between ash particles ejected from the volcano. A Japanese folk belief says that earthquakes will follow lightning from a clear sky. Since earthquakes are known to be associated with Sakurajima and other volcanoes, such lightning may indeed be a harbinger of earthquakes. Elsewhere, "dry lightning" sometimes occurs when moisture-laden air blows over warmer air. Photo: Tsuyoshi Nishiinoue*

RIGHT: *The eruption of Mt. Kilauea, Hawaii, in 1984 pumped clouds of ash into the moisture-laden atmosphere. Heating of the air above the volcano created strong updrafts similar to those of a severe thunderstorm. Static electricity produced by colliding ash particles created this stunning display of lightning. Photo: E. R. Degginger/ Dembinsky Photo Associates*

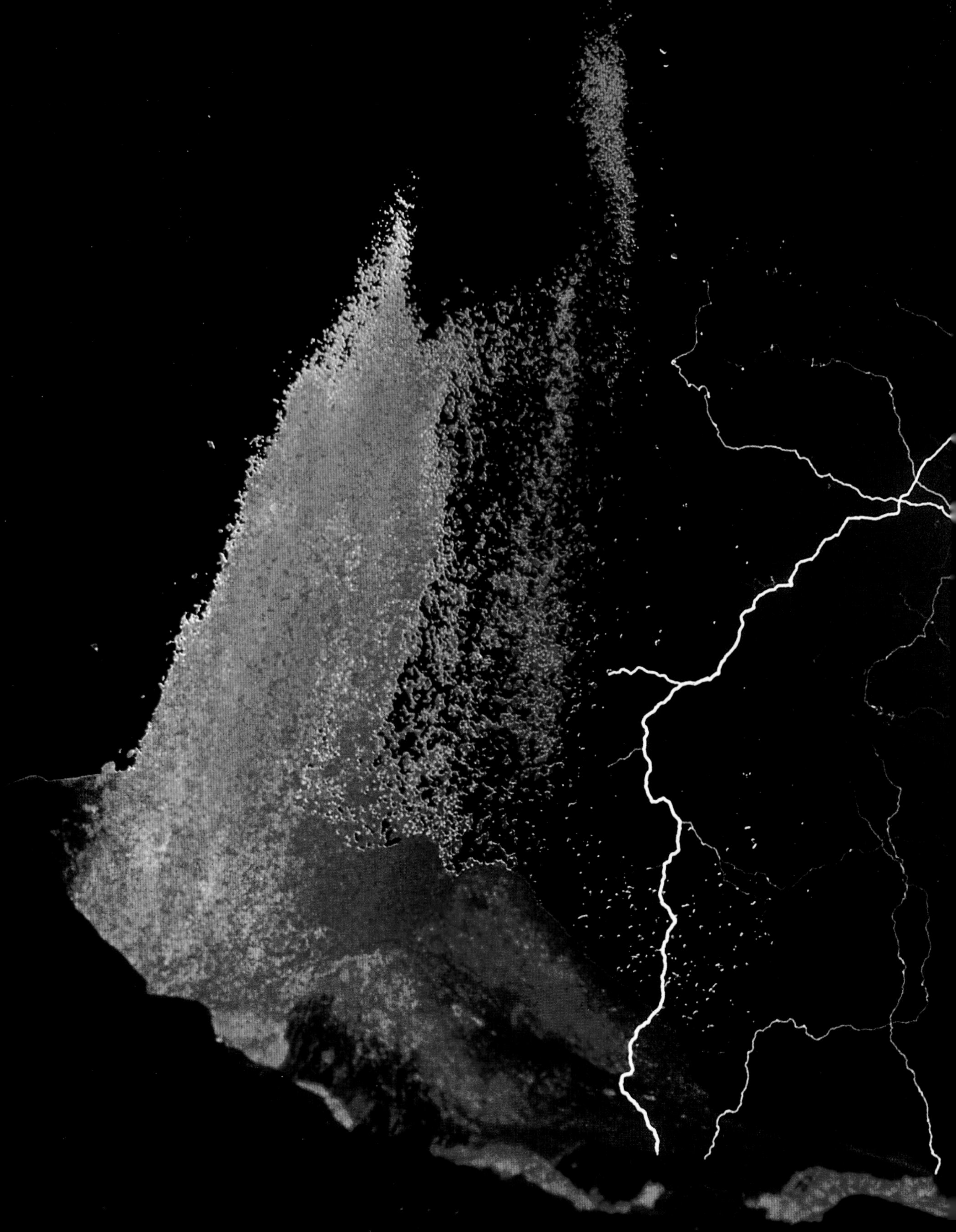

CONTENTS

FOREWORD

PRECEDING PAGE 12: *May 5, 1993, was a perfect day in tornado alley unless you were caught in the path of this F3 tornado in Hansford County, Texas, or any of the other six tornadoes this storm spawned as it traveled across the Oklahoma Panhandle and into the plains of Kansas. The tornado was on the ground for 37 minutes and destroyed a farm before roping out 4.5 miles east of Guymon, Oklahoma. Photo: A. T. Willett*

LEFT: *Layered lenticular clouds, formed by moisture-laden air flowing over the Sierra Nevada in the background, float over strange tufa towers on Mono Lake in the Mono Basin National Forest, California. Formed from calcium carbonate deposits from volcanically heated vents at the bottom of the lake, the towers emerged as the water from the lake was drained over several decades to quench the thirst of Los Angeles, more than 200 miles away. The lake is now being preserved and the water will someday, once again, cover these monoliths. Photo: Fred Hirschmann*

Earth has weather—and life. Other planets may be hotter and drier, or have higher mountains and stronger storms, but only on earth do the elements of air, water, ice, clouds, vegetation and geology combine to give us the weather and climate that support and threaten life. With striking images and concise text, the contributors of the *Weather Guide Calendar* have brought together in *Weather: The Ultimate Book of Meteorological Events* a stunning vision of this dynamic physical environment.

The images of weather, water, light, wind, and climate change give the reader a proper sense of awe about the powerful forces that shape our world. Supercell storms with fearsome winds and heavy rainfall are pictured here with their accompanying tornadoes that add destructive damage. From dust devils in the Kalahari Desert and killer tornadoes in Kansas, to melting glaciers and ice caps, the images remind us of both our vulnerability and how we are changing earth.

Earth's weather is driven by the heat from the sun, which is absorbed by the land and by the ocean which heats the atmosphere from below. Warm air near the surface rises to form clouds, winds, and much of what we see as weather. Over the ocean, the winds drive the ocean currents and in turn are affected by the heat transferred from the ocean to the atmosphere. It's all a kind of controlled chaos. The complex system of air, water, land and ecosystems is acting according to the laws of motion and of thermodynamics. The energy for all of this comes from the sun. Once the system is set in motion, it evolves to what we call weather. Never has chaos been so beautiful.

One thing that should be very clear from this book—weather has many faces and is hard to predict. This complex system is only slowly yielding its secrets of predictability to the great scientific minds and computers of our time. Today, we know that unpredictable turbulence makes weather forecasts simply not possible beyond two weeks. As computers get bigger, more data come in from radars and satellites, and scientists understand more about the whole system, the predictions will improve.

Will we ever get to a two-week forecast? It's possible, but will take a long time. We see incremental improvement as computers get better, we have more observations, and satellites monitor the earth. We have new systems for forecasting river levels, and the entire United States is covered by modern Doppler radar systems that track every storm. We have the best weather forecasting in the world, but there is still much room for improvement.

I can say that the band of weather forecasters are among the most dedicated people in the world. These are committed scientists who love their work. When I was the head of the U.S. National Oceanic and Atmospheric Administration, I met many forecasters in government and in the private sector who spent the day at their job making and improving forecasts, and then when they went home, joined computer chat rooms and blogs to talk more about weather—how fascinating it is and how to improve the forecasts. And the amateur weather spotters are an important part of this system as well.

Weather and climate have always been with us, and have always been changing. We know, for example, that hurricanes have stormed the coasts long before any humans were there, and El Niños have been recorded for millennia. What we see today is like what has happened in the past. Sometimes it was less stormy, sometimes more. But now this is all changing as humans add so-called "greenhouse gases" like carbon dioxide to the atmosphere by burning fossil fuels. These newly added greenhouse gases are causing the earth to warm, at a rate that is unprecedented in history.

Will this global warming change the weather? The average temperature of the oceans and earth's atmosphere is increasing, and we are now seeing dramatic changes in the polar regions, where the ice cap is retreating at an alarming rate. The ice cap melt water is raising sea level, threatening low-lying coastal areas worldwide. It is likely we can expect more heat waves and other extreme weather events in the near future. *Weather: The Ultimate Book of Meteorological Events* truly reflects the engrossing nature of a topic of critical importance to all of us.

—Dr. D. James Baker, Philadelphia, October 2007

Scattered air-mass thunderstorms form along the coast of Australia's Northern Territory. This shot, from an altitude of 190 nautical miles, was taken by the Shuttle Columbia crew in November 1996. Photo: NASA

SIGNIFICANT EVENTS IN METEOROLOGICAL HISTORY

600 BC
Assyrians record weather lore in cuneiform writing.

570 BC
Anaximenes of Miletus introduces the concepts of condensation and rarefaction and suggests that wind, rain, and clouds are formed by air in different states.

400 BC
The Greek physician and scientist Hippocrates writes the treatise *Airs, Waters, and Places*, linking health to climate.

340 BC
Aristotle writes the *Meteorologica*, part of which is devoted to atmospheric phenomena and from which we get the word "meteorology"—the study of earth's atmosphere and the variations in atmospheric conditions that produce weather.

48 BC
Andronicus, a Greek astronomer, constructs the Tower of the Winds in Athens. It is the earliest known weather station—and it is topped by the earliest known weather vane.

AD 100
Hero of Alexandria experiments with air pressure and discovers that air expands when heated.

1450
Nicholas of Cusa invents a type of hygrometer to measure humidity, using the fact that certain organic materials such as hair increase in length with increasing atmospheric humidity.

1450
Leon Battista Alberti invents an anemometer to measure wind speed. His device measures the effect of wind on a flat plate of metal hinged to move against a protractor-like scale.

1597
Italian astronomer Galileo invents the first known thermometer, called the thermoscope, using ideas first investigated by the Greek Philo in the third century BC.

1643
Evangelista Torricelli and Viviani of Italy, pupils of Galileo, construct the first mercury-based barometer.

1650
The Climate Academy in Florence, Italy, is created to develop new instruments and record weather events.

1686
Edmund Halley, now best known for the prediction of the return of the comet that bears his name, identifies pressure differences over land and water as the driving force behind the winds. He also produces a meteorological map of the trade winds.

1714
Gabriel D. Fahrenheit, a German physicist, builds a mercury thermometer (having made one with alcohol in 1709) with the temperature scale that now bears his name.

1728
Swedish explorer Vitus Bering, employed by the government of Russia, begins a series of expeditions to Siberia and the North Pacific. The Russian Academy of Sciences instructed Bering to take measurements of temperature and barometric pressure, as well as to make observations of clouds, thunderstorms, and related natural phenomena. Bering established a series of observation stations throughout northern Siberia, providing the first scientific knowledge about the vast region.

1735
George Hadley introduces the "Hadley cell," which describes a wind circulation pattern in the equatorial regions that affects global winds.

1752
Benjamin Franklin performs his famed kite experiment to prove that lightning is electrical.

1784
Benjamin Franklin links recent cool, hazy weather to the eruptions of volcanoes in Iceland.

1807
President Thomas Jefferson orders the creation of the Coast Survey, the federal government's first scientific agency. Among its responsibilities are keeping and analyzing weather records.

1814
In 1814, Surgeon General James Tilton issues a general order directing all surgeons in the U.S. Army to take daily weather observations.

1827
French mathematician Jean Baptiste Fourier hypothesizes about human effects on earth's weather.

1842
Elias Loomis produces the first weather map of a storm that occurred over the eastern U.S. in 1836.

1844
American inventor Samuel F. B. Morse perfects the telegraph, which enables meteorologists to send weather observations from one city to another quickly.

1848
The *Washington Evening Post* publishes weather reports from a network of 200 weather observers set up by the Smithsonian Institution.

1849
Joseph Henry, the secretary of the Smithsonian Institution in Washington, D.C., receives the first weather report sent by telegraph in the U.S. This same year the Smithsonian establishes an expanded observation network around the nation, transmitting data by telegraph.

1860
Great Britain follows France's lead and begins a telegraphic weather service.

1860
According to Weather Service records, limited weather data being recorded and reported from 500 stations is interrupted by the outbreak of the Civil War.

1869
Cleveland Abbe issues the first weather bulletins from a Cincinnati observatory.

1870s
Fishermen off the coast of South America coin the phrase "El Niño" to describe the appearance of unusually warm water in the Pacific Ocean around Christmastime. (The baby Jesus is called El Niño in Spanish.) Born in the far western Pacific, the El Niño system brings intense rains to the Americas.

1870
The U.S. National Weather Service forms under the auspices of the Army. President Ulysses S. Grant signs a joint resolution of Congress authorizing the Secretary of War to establish a weather service within the Army. The observations are made at 22 locations by the Army Signal Corps, and the word "forecast" becomes established.

1871
Canada establishes its weather service.

1890
Congress organizes an agency called the Weather Bureau. Its name is changed to the National Weather Service in 1967.

1890s
Data from balloons and kites provide the first soundings from the upper layers of atmosphere, leading to the discovery of the stratosphere.

1891
The Secretary of Agriculture orders the Weather Bureau to conduct rainmaking experiments by detonating explosive charges in the air.

1898
President William McKinley orders the Weather Bureau to establish a hurricane-warning network.

1902
Teisserenc de Bort divides the atmosphere into the troposphere and stratosphere.

1903
Orville Wright consults with the Weather Bureau before making his historic first powered flight at Kitty Hawk, North Carolina. Within the next year airplanes are used to collect atmospheric data.

1904
German engineer Christian Hulsmeyer describes a telemobiloskop, which he envisions as a method of using radio waves to detect ships at sea and prevent collisions. His dream was not realized, however, until further work on radar (Radio Detection and Ranging) in the 1920s and more complete development during World War II.

1920s
British mathematician Lewis Fry Richardson, believing that because the behavior of the atmosphere follows the laws of physics, mathematics can be applied to predict weather, develops calculations that applied those laws to the varying conditions of the atmosphere. His calculations take so much time to perform, however, that the weather conditions pass before a forecast can be prepared.

1924
British scientist Sir Edward Victor Appleton bounces radio waves off the ionosphere and uses the reflected waves to deduce the height of that layer—a successful application of the principle of radar.

1927
The first radiosonde (originally "Rawinsonde" or "radio-wind-sounding device") is sent aloft by balloon to radio back information from high in the atmosphere.

1928
Teletype replaces the telegraph for weather communications.

1935
Robert Watson Watt invents radar to track aircraft, based on experiments in the 1920s of bouncing radio waves off the ionosphere.

1937
The first official Weather Bureau balloon-launched radiosonde soundings are made at East Boston, Massachusetts. This program draws aircraft soundings to an end because balloons can reach heights of 50,000 feet.

1940s
Radar, developed for military purposes, turns out to be equally valuable in tracking weather.

1940
The first practical electronic digital computers are developed. Led by John von Neumann, a group of American meteorologists and mathematicians begins formulating equations that will enable computers to predict the weather.

1942
The Weather Bureau inherits 25 surplus radars from the Navy to be used in meteorological research.

1948
U.S. Air Force Weather Services issues the first tornado warnings for military installations.

1948
The first primitive weather forecasts are made (based on numerical data) by the electronic brain, ENIAC (Electronic Numerical Integrator and Computer).

1950
The first successful weather forecast prepared by computer is announced.

1950
Carl-Gustav Rossby discovers the global, high-altitude current of air now known as the jet stream.

1954
The U.S. Air Force debuts the first radar system specifically designed for meteorology, the AN/CPS-9.

1959
The United States launches the first satellite to send weather information back to earth.

1959
The first WSR-57 (non-Doppler) weather surveillance radar is installed at the Miami Hurricane Center.

1960
Tiros I, the first weather satellite equipped with a camera, is launched into orbit by NASA.

1963
Edward Lorentz publishes a paper entitled "Deterministic Nonperiodic Flow" showing how very small changes in meteorological conditions can lead to large and unpredictable effects ("chaos"). Today many physical systems including weather are thought to be affected or controlled by chaotic processes, which lead to unpredictability.

1967
The U.S. Weather Bureau is renamed the National Weather Service (NWS).

1970
The National Oceanic and Atmospheric Administration (NOAA) is founded as an independent federal agency. Its authorization is the Marine Resources and Engineering Development Act of June 1966, passed following congressional concern that 45 percent of the people of the United States were living in coastal areas and required better information about impending storms.

1971
The first Doppler Weather Radar goes into operation at the National Severe Storms Laboratory (NSSL) in Norman, Oklahoma. Previous Doppler radars had been used for commercial purposes, and had been experimented with since the 1960s. Doppler radars have the advantage of providing not merely distance and direction of storms and precipitation, but also wind speed and motion within the atmosphere. This eventually led to the development of the next generation of weather radar (formerly called NEXRAD) now embodied in the WSR-88D radar network.

1973
The National Weather Service purchases its first Doppler radar, the WSR-74C, now superseded by the more advanced WSR-88D version.

1974
NASA launches the first prototype GOES (Geostationary Operational Environmental Satellite) satellite, called the Synchronous Meteorological Satellite (SMS-1), to test advanced imaging in visible and infrared light.

1975

The first GOES launched satellite, GOES-1, goes into orbit to provide state-of-the-art continuous monitoring of earth's atmosphere and surface over a large region of the Western Hemisphere. GOES satellites monitor potential severe weather conditions, such as tornadoes, flash floods, hail storms, and hurricanes.

1976

The first weather report from the surface of another planet, Mars, comes from NASA's Viking spacecraft. The average atmospheric pressure is about 0.007 millibars, or less than 1/100th of that on Earth. The average summertime temperature is –72° Fahrenheit, or –58° Celsius. Wind gusts of up to 65 kilometers per hour (42 miles per hour) are recorded. Extension of the mission allows weather-data gathering into the Martian winter, when temperatures drop to as low as –171° Fahrenheit (–113° Celsius).

1979

The first World Climate Conference is convened by the World Meteorological Organization. The conference will lead to international attention to weather matters, which will be of particular importance in the early 21st century. Among other things, the WMO establishes that the names of hurricanes would be both male and female. When the National Weather Service began naming hurricanes in 1953, it used only female names.

1982

The first Synoptic Flow experiment is conducted, using dropsondes around Hurricane Debby to help define the large-scale atmospheric winds that steer such storms.

1990

The National Weather Services contracts for the first WSR-88D Doppler radar (NEXRAD, or Next Generation Radar). NEXRAD radars are fully automated and, unlike ordinary radar, can probe the interior workings of storms, allowing imaging of the structure and wind direction and speed. They also can detect dangerous wind shear, which makes them especially important near airports.

1992

Greenland ice cores provide evidence that climate change during the last Ice Age was often dramatic and very quick, going from clement to icy in as little as a year.

1997

NASA launches the GOES-10 satellite. GOES-10 (originally designated GOES-K) is the latest in the GOES series. This 4,600-pound spacecraft continuously provides data and images on cloud cover, storm activity and other atmospheric variables over North America. It is stationed over the Equator directly south of Denver (105° west). It also serves to relay weather information from ground-based stations, as well as to relay distress signals from individuals, aircraft, and ships to search and rescue stations. GOES-10 replaced GOES-9, which had been experiencing overheating in some of its components. Additional GOES launches are planned.

2000

Computer and other technologies improve weather-forecasting abilities to the point that five-day forecasts for the weather over North America and Europe are as accurate as three-day forecasts were a quarter-century before. Regional climate patterns can now be predicted with a high rate of accuracy months in advance.

2001

Established in 1988, the Intergovernmental Panel on Climate Change (IPCC) begins issuing scientific reports confirming the connection between human activities, principally the generation of greenhouse gases, and climate change. CAMEX4, a NASA experiment run in conjunction with NOAA's Hurricane Field Program, collects detailed data on hurricanes Erin, Gabrielle, and Humberto and tropical storm Chantal.

2005

Hurricane Katrina devastates the Gulf Coast, a storm so powerful that henceforth the name Katrina is retired from the World Meteorological Organization's list of approved names for hurricanes. Though accurately predicted, Katrina finds federal agencies ill prepared to respond.

2006

NASA's African Monsoon Multidisciplinary Analyses (AMMA) experiment examines the wind regimes over western Africa and their role in generating disturbances over the Atlantic. The experiment uses long-range aircraft and the Cloud-Aerosol Lidar and Infrared Pathfinder Satellite Observation (CALIPSO) weather satellite, launched on April 28 of that year.

2007

Former American Vice President Al Gore's film *An Inconvenient Truth* wins an Academy Award for best documentary—the first film about weather and climate to take that coveted prize.

TOP U.S. WEATHER EVENTS

TOP 10 EVENTS OF THE 20TH CENTURY

WeatherWise *magazine (Heldref Publications) compiled this list in 1999 by polling some of its regular contributors. The final entrants were selected based on their magnitude and scale as well as their importance in the development of meteorological science. —Kim Long*

1. Dust Bowl, 1930s

In the 1930s, a drought plagued much of the country, resulting in a massive "Dust Bowl." The drought lasted for a decade, with the heaviest effects concentrated in Oklahoma, Texas, Colorado, and Kansas. The killer drought peaked in July of 1934, but one of its most memorable events was on April 14, 1935, known as "Black Sunday," when a massive cloud of dust and blowing topsoil blanketed most of five states.

2. Super Tornado Outbreak, 1974

On April 3 and 4, 1974, a widespread weather system produced 48 tornadoes, with 30 of these measured at F4 or F5 intensity. The multi-tornado event killed 300 people across 13 states and set a record for the most twisters in the widest geographic area.

3. Galveston Hurricane, 1900

On Saturday, September 8, 1900, between 6,000 and 12,000 people were killed when a hurricane struck Galveston Island, Texas. The powerful storm—known as the Galveston Hurricane (because the modern practice of naming hurricanes did not begin until 1960)—caused more than $30 million in damages with most of the death and destruction of property due to a huge mass of water surging in from the ocean.

4. Superstorm, 1993

From March 12 through 15, 1993, a huge storm system produced heavy snow, rain, high winds, and multiple tornadoes throughout the East Coast. Several hundred deaths resulted from the storm, which produced one of the widest-ranging snowfalls in the century.

5. Tri-State Tornado, 1925

On March 18, 1925, a killer tornado produced the longest-known track on the ground—219 miles (352 km)—spanning Missouri, Illinois, and Indiana. Close to 700 people were killed—a record-setting 234 in one city, Murphysboro, Illinois—during the 3½ hours this twister was on the ground and more than $16 million in damages resulted.

6. Hurricane Andrew, 1992

On August 24, 1992, a Category 4 hurricane slammed into Miami, Florida. It caused about 50 deaths and $25 billion in damages, with an estimated 250,000 people left homeless.

7. Hurricane Camille, 1969

On the night of August 17, 1969, a Category 5 hurricane swept in from the Gulf of Mexico to batter the Mississippi Coast. Only one other Category 5 hurricane made landfall during this century. Camille produced winds up to 172 mph (277 kmph) and continued wreaking havoc as it swept inland, producing record rainfall and flooding in Virginia.

8. Mississippi Flood, 1927

In the spring of 1927, torrential rains throughout the Mississippi Valley produced widespread flooding. An estimated $284 million in damages resulted, with 215 deaths reported; almost 1 million people were evacuated. But the death toll may have been much higher, perhaps in the thousands, because of under-counting the number of African Americans caught in the flood.

9. El Niño, 1997–1998

The appearance of warmer waters off the western coast of South America triggered worldwide changes in climate and weather. Although not as strong as the El Niño of 1982–1983, this event is credited as the first to be forecast in advance by scientists.

10. New England Hurricane, 1938

On September 21, 1938, a large hurricane struck New England without warning. Marked by winds of up to 186 mph (299 kmph), the storm followed an unusual track, moving north. Damage was extensive from the wind, and heavy flooding also caused widespread losses; the death toll was about 600 people.

BELOW: *Flooded neighborhoods are part of the devastating aftermath of Hurricane Katrina. Photo: F. Jocelyn Augustino/Weatherstock*

TOP 5 EVENTS, 2000–2007

When it rains, the adage tells us, it pours. And when it doesn't rain, it does whatever the reverse of pouring might be. America's weather in the first years of the 21st century was characterized by hurricanes, floods, tornadoes, and severe drought. Here are the top five events for the century to date (that is, mid-2008, when this book went to press). And as for what the rest of the decade and the coming years of the 21st century will bring? Stay tuned. —Gregory McNamee

1. Hurricane Katrina, 2005

Born as a tropical depression off West Africa, Katrina reached Florida on August 25, 2005, as a relatively weak Category 1 storm, then gathered strength when it crossed the peninsula and found a new source of energy in the warm waters of the Gulf of Mexico. When it hit the Gulf Coast on August 29, it was still only a Category 3 storm, with winds of about 130 mph. Yet accidents of climate and history combined to make it a perfect storm: the powerful surge that accompanied it inundated New Orleans and caused flooding as far inland as Indiana, while wind and debris ruined countless structures from northern Florida to eastern Texas. In the end, the storm cost more than 1,800 lives and $125 billion.

2. Hurricanes Rita and Wilma, 2005

A Gulf Coast—and a nation—already rattled by Katrina suffered two major hurricanes in the immediate aftermath. Rita arrived in September 2005. The Category 3 hurricane hit the border of Texas and Louisiana, a region of critical importance to the U.S. petroleum industry, and caused damage and flooding. Before it landed, Rita reached the third lowest pressure (897 mb) ever recorded in the Atlantic basin, which means, thankfully, that had the storm had more energy it could have been more destructive than Katrina. The cost was $16 billion and an estimated 119 deaths. Wilma hit southern Florida in October 2005, having declined from a Category 5 to a Category 3 storm. Wilma marked the lowest pressure (882 mb) ever recorded in the Atlantic basin. It, too, cost $16 billion, but only 35 lives were lost.

3. Widespread Drought, 2006

In the spring and summer of 2006, widespread drought, along with unusually high temperatures, affected crops throughout the Great Plains, but also touched major portions of California, the Southwest, and the Pacific Northwest. Damages have been estimated at more than $6 billion, but, thanks to good emergency warning measures, no more than the statistically normal number of heat-related deaths were reported.

4. Super Tornado Outbreak, 2007

A severe weather front spawned dozens of tornadoes over the Great Plains on May 4, 2007, including one that clocked wind speeds of more than 205 mph and obliterated the small town of Greensburg, Kansas. Fortunately, only 11 people died in the town and surrounding area. The same storm system spawned an astonishing 91 tornadoes the next day.

5. Three Major Hurricanes, 2004

Three major hurricanes hit Florida and the Gulf Coast in September 2004. Jeanne, a Category 3 hurricane that made landfall in east-central Florida, caused considerable wind, storm surge, and flooding damage and cost $7 billion and 28 lives. Ivan, another Category 3 storm, hit the coast of Alabama and caused flooding as far inland as New York; its costs were more than $14 billion and at least 57 deaths. Frances, again arriving in east-central Florida, caused flooding and wind damage; in the end, it yielded $9 billion in damage and at least 48 deaths.

FORECASTING

100 YEARS OF WEATHER OBSERVING

A century ago, in the act establishing the Weather Bureau within the Department of Agriculture, Congress directed the bureau to "record the climate of the United States." Through the intervening years, a dedicated group of unpaid cooperative weather observers has provided weather and climate records for more than 25,000 locations in the United States. Using instruments furnished and maintained by the Weather Bureau (now the National Weather Service), and following procedures and conventions established for that purpose, this network of volunteers has produced a treasure trove of information about our nation's many and varied climates.

At present, observers at nearly 10,000 sites donate over a million hours of their time each year to participate in this effort. Observers are located at universities, homes, farms, municipal facilities, utilities, dams, parks, refuges, and radio and television stations. Many of the individual observers have made daily measurements for 30, 40, or 50 years and more. The longest participation on record by one person was by Mr. Edward G. Stoll, who took observations for 76 years in Arapahoe, Nebraska. Successive family generations have provided very long records, with several reaching a hundred years or more.

The information gathered has been widely used for sectors of the economy as diverse as agriculture, business, commerce, industry, engineering, and aviation. These records continue to acquire greater value with the passage of time. As concern increases about the effect of human activities on global climate, these unique and irreplaceable observations will be vital for the detection and description of any changes in climate, which may occur at local and regional levels.

More than half a century ago, a group of outstanding scientists reported to President Franklin D. Roosevelt that the work of the cooperative weather observers "is one of the most extraordinary services ever developed anywhere and probably nets the public more per dollar expended than any other government service in the world." This statement remains true today.

—National Climatic Data Center

Ancient philosophers speculated the world was composed of four humors: earth, air, fire, and water. Modern physics describes a more diverse universe, yet what we experience as weather and climate essentially emerge from the mixture of gases called air, plus water in its three phases (gas, liquid, and solid), earth (the tiny specks of matter upon which all rain and snow must condense), and the atmospheric motions driven by the unequal distribution of heat received from the solar fire.

Our atmosphere, proportionally thinner than the skin of an apple, is home to an endless parade of swirls and eddies, calms and gales, droughts and deluges, searing heat and numbing cold that affect humanity in myriad ways. The peaceful pleasure of enjoying a sea breeze on the shore while the reddened sun sets can overnight be replaced by the terror of the banshee howl of winds and an inexorable tropical storm surge drowning a coastal village.

Forecasting nature's moods was once the realm of sorcerers, astrologers, and early astronomers. Those who foretold a pending eclipse were rewarded for serving the local ruler's political ends. Those who failed sometimes met tragic fates. The first scientific attempts at weather forecasting in 1860s England saw Robert Fitzroy bravely striving to save lives and livelihoods by teasing the next day's weather from scattered observations telegraphed to his London office. Yet even he paid the price of frequent failure, as public derision was a likely factor in him taking his life in 1865.

PRECEDING PAGE 25: *The monsoon season usually makes it way into southern Arizona in late summer when southerly winds bring humid air from the Sea of Cortez and the Gulf of Mexico to meet Arizona's hot, low-pressure air. The convergence of these two air masses is often violent, resulting in intense thunderstorms that persist well into evening hours. Photo: Adam Block*

LEFT: *Nature sometimes provides seemingly incongruous combinations, such as snow on these Saguaro cacti in this canyon of the Santa Catalina Mountains near Tucson, Arizona. The combination of winter precipitation and low temperatures that formed this scene is rare in the Sonoran Desert of southern Arizona. Photo: Jack Dykinga*

By 1870, the first U.S. public storm warnings were issued, and in 1890, the formation began of what is now the modern National Weather Service. Progress was steady, but slow. The atmosphere—and meteorology—is international in scope. During World War I, Scandinavian scientists' detailed analyses of Nordic weather maps discovered the warm and cold fronts now marching familiarly across our TV weather maps. These pioneers were specialists in fluid dynamics who noted that air is, in fact, a fluid.

The same equations of fluid behavior could be applied to prognosticate fronts, cyclones, and anticyclones. In the 1920s, Lewis Richardson proclaimed a legion of 10,000 accountants armed with calculators could solve those equations before tomorrow's weather occurred. This absurd notion suddenly became practical with the advent of the digital computer in the 1950s. Soon a *New York Times* story claimed future computers would someday generate personalized forecasts down to the street-corner level. Indeed, today's supercomputers do transmit surprisingly accurate forecasts to your cell phone, at least with zip code precision, and perhaps soon, at the level of that same street corner.

Once the butt of jokes for their inaccuracy, forecasters have developed sufficient skill that people are surprised when a major storm prediction is busted. They save thousands of lives annually, and though economic losses are still considerable, the savings are far greater in the 30 percent of the massive U.S. economy which is weather sensitive. Whether the prognosis is for massive waves to delight surfers, swells to be avoided by giant container vessels, or jet-stream tailwinds to be ridden by jumbo jets, modern weather forecasting is a triumph of the tenacity and courage of legions of scientists. Robert Fitzroy would be pleased.

—Walter Lyons, CCM

LEFT: *A supercell thunderstorm produced this cloud-to-ground bolt of lightning at sunset near Medicine Lodge, Kansas. The severe storm dropped baseball-sized hail before losing its flying saucer-like shape. Supercells are generated by rotating, rising columns of air called "mesocyclones" that can give parts of the clouds a corkscrew shape. The warm air rises at speeds up to 170 mph accounting for the hard, cauliflower-like appearance of supercells, which can last for hours and move great distances. Photo: Jim Reed/Getty Images*

ABOVE: *This F3 tornado, with winds up to 200 mph, touched down in rural South Dakota, near Aberdeen. Not more than 5 seconds after this shot was taken, the tornado destroyed the farm in the foreground. Photo Courtesy of NOAA*

LEFT: *It is nearly a cliché that tornadoes target trailer parks. While there is no deliberate intent on the part of nature, mobile homes certainly are at risk. This twister is in the process of demolishing a mobile home near Spring Valley, Texas. It was the first of at least eight tornadoes that menaced central Texas on this day (May 27, 1997), which killed 27 people in the town of Jarrell. Photo: Lon Curtis*

RIGHT: *This classic, anvil-shaped, single-cell thunderstorm dumped rain and threw lightning over Tuscon, Arizona, in September, 1983. The silhouettes seen in the foreground are two people who were enjoying the show at twilight. Photo: A. T. Willett*

The highest average annual rainfall in the United States occurs in Mt. Waialeale on the island of Kauai. An average of 460 inches (1,168 cm) of rain falls every year.

NORTH AMERICAN STORM TRACKS

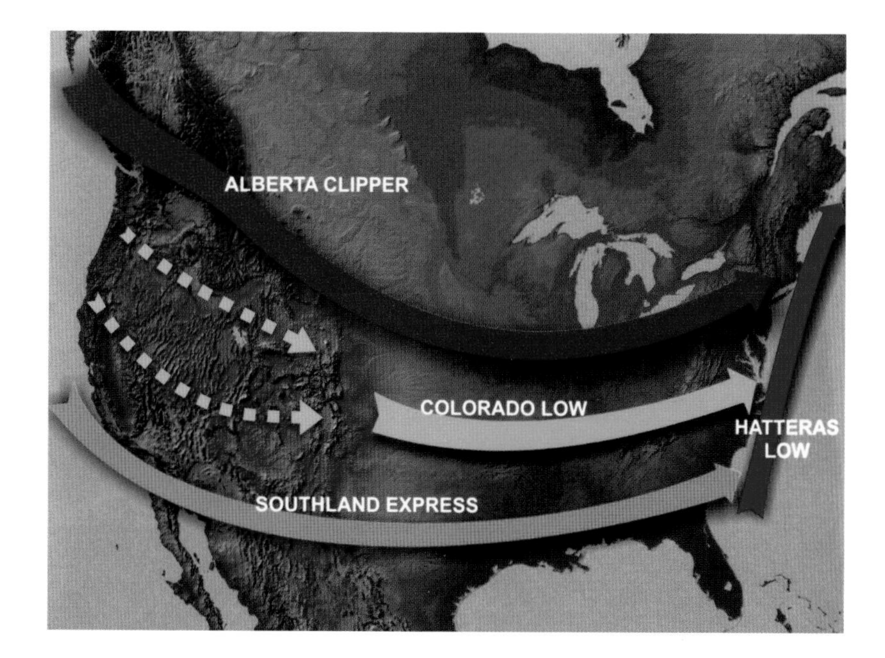

With a canopy of dark clouds overhead, the South Rim of the Grand Canyon provided a magnificent target as it was by the crackle of intense cloud-to-ground lightning. Monsoon rains, always needed in this arid part of the Colorado Plateau, followed this shadow of lightning captured from Point Sublime on the North Rim. Photo: Dick Dietrich/Dietrich Stock Photos

In some parts of the country, it seems that storms follow the same paths most of the time, almost as if they were following an invisible itinerary. This is not a case of flaky observation; many local storms, particularly in hilly or mountainous geography, are spawned and directed in part from the influence of terrain. In the Front Range of the Rocky Mountains of Colorado, for example, summer thunderstorms are often prevalent along certain northeasterly paths. A local weather forecaster was once made notorious for predicting a successful—as in rain-free—outdoor event, but only if it were held at one end of a city park and not the other.

Larger storm systems and fronts are also influenced by geography. Driven by the conditions associated with large bodies of water, ocean currents, mountain barriers, and seasonal jet stream patterns, the United States can be said to have four major storm tracks. These may vary from year to year—and from season to season—but they represent a general, known pattern upon which regional and local weather forecasters may rely, at least for historical comparison.

In the north, the Alberta clipper is a well-recognized presence, often identified in media headlines attached to weather changes. Not surprisingly, this storm track gets its name from the Canadian province of Alberta, where it seems to originate. However, the clipper actually arises in the northern Pacific Ocean and, linked to the jet stream, drops its initial load of moisture as it shifts up and over the mountainous barrier it encounters in coastal British Columbia. Flowing eastward over additional mountains in Alberta, the now-drier clipper typically collides with the cold air mass often stationed—and stationary—over the western Canadian plains. High-pressure systems centered to the north can push the flowing clipper southward, producing major winter storms in the northern United States and Great Plains; otherwise, it heads more or less easterly across Canada.

Although most Alberta clippers are accompanied by some moisture, they are not usually known for heavy snowfall unless they interact with other storms pulling moisture up from the Gulf of Mexico. Characteristics of a typical Alberta clipper include harsh winds and sharp drops in temperature, as much as

Many storms congregated in the sparsely populated north Texas Panhandle on May 29, 1991 to produce spectacular shows of lightning. Several of the storms were supercells, including this one, and produced severe weather consisting of mostly large hail and strong winds. At sunset, this storm appeared to be raining lightning, dwarfing the windmill below it. Photo: Charles A. Doswell III

30 degrees Fahrenheit (16 degrees Celsius) in less than 12 hours. Clippers are common throughout the year, but they are coldest and most damaging in winter months. They can come at a rapid pace, one or two a week for several weeks in a row.

In the center part of the country, the major storm system is the Colorado low, which appears to pop up in southeastern Colorado. It is sometimes referred to as the Colorado-Trinidad low, because an apparent point of origin may be the town of Trinidad, near the southern border of the state. Such a storm track may also originate even farther south, in northern New Mexico. The Colorado low typically starts as a low-pressure mass and moves east to northeast, sometimes affecting cities as far north as Chicago. Usually it produces heavy winter precipitation as far east as the Atlantic coast. In the winter, Colorado lows can be pushed down by the jet stream or blocking air masses to the north, and can bring icy conditions and heavy snow as far south as Texas, where these storms are traditionally called blue northers.

Although most pronounced during winter months, when they can produce prodigious, wide-ranging blizzards—Colorado lows produce most of the Midwest's snow—this storm track also sweeps through the Midwest in the summer, pushing thunderstorms and widespread rain ahead of a cold front. At an extreme, a summer Colorado low can bring heavy thunderstorms that linger for several hours over an affected area.

To the south, the Southland express produces the most recognizable storm track as it moves across the country. Originating in the Pacific Ocean waters off northern California, it sweeps down and crosses land in the southern part of the state, driving eastward toward Florida and the Atlantic. As it crosses the Gulf states, it can be affected by moisture from the Gulf of Mexico, as well as fronts to the north. The Southland express is also referred to as a Gulf coaster, particularly when it pushes low across the Gulf states and Gulf of Mexico. When it blows through, it is known for severe weather, particularly heavy rainfall—snow when freezing temperatures coincide—and it can push upward into the Midwest and East Coast states.

The fourth major storm track affecting the United States is called the Hatteras low. Linked to the area around Cape Hatteras, off the North Carolina coast, a typical outcome for this system is to push up along the Atlantic coast; in New England, the Hatteras low is also known as a nor'easter. This name originated in colonial days, when weather observers named storms for the direction from which the wind was blowing. Even though the Hatteras low is moving to, not coming from, the northeast, its counterclockwise movement makes it appear just the opposite.

Capable of producing some of the most intense storm conditions, the Hatteras low is sometimes referred to as a "bomb" cyclone. Cyclone because the air moves counterclockwise (like all North American storm tracks); bomb because such a storm can develop very quickly. The warm water of the Gulf Stream lies next to the colder waters of the Atlantic coast, a situation ripe for extreme weather and generating heavy rain—or snow in wintry conditions—as far north as the Maritime provinces in Canada. At their worst, Hatteras lows produce extreme weather both on land and at sea. "The Perfect Storm" of best-seller book fame (as well as the blockbuster movie version) was a Hatteras low.

These four major tracks are mostly independent of one another but, under some circumstances, can combine or collide, dramatically increasing their stormy potential. In November, for example, the Great Lakes region can provide a target zone where an Alberta clipper and a Colorado low converge, their effects made more severe by the plentiful supply of moisture—and source of heat—represented by the Great Lakes. At least 25 "killer storms" have been recorded on the Great Lakes in November, producing more deaths from shipwrecks in that month than at any other time of year.

—Kim Long

TORNADO FORECASTING:
A BRIEF HISTORY

Although Benjamin Franklin realized as early as 1743 that storms in the United States generally moved from west to east or southwest to northeast, the lack of rapid communications at the time hindered warning regions in a storm's path. It was not until a century later, when the first commercial telegraph line was opened on April 1, 1845, that forewarning communities of approaching severe storms and tornadoes was possible.

Joseph Henry, America's premier physicist and the first director of the Smithsonian Institution, proposed in a letter to the Smithsonian's regents dated December 8, 1847, that the institution "organize a system of observation which shall extend as far as possible over the North American continent." The following year Henry initiated a volunteer weather observation program.

By 1860, the Smithsonian volunteer network encompassed more than 500 reporting stations. A series of devastating spring and summer tornadoes that year prompted Henry to request information from eyewitnesses. In 1862 the Smithsonian distributed circulars to the public warning of the dangers posed by tornadoes and asking for continued reports and data on these storms. The public's response was so great that in 1872 the institution issued a four-page pamphlet listing the questions observers should attempt to answer when reporting a tornado, including the date, location, length and width of path, direction of speed of movement, color of sky, and shape of funnel.

Accounts of intense storms might have occasionally appeared in the *Monthly Weather Review* or the *Monthly Weather Summary of States,* but meteorologists appeared to have reached a general consensus that forecasting tornadoes would do more harm than good. Continuing the precedent set by the Signal Corps, the Weather Bureau Stations Regulations of 1905 contained the statement, "Forecasts of tornadoes are prohibited." When conditions were favorable for tornado formation, district forecasters could use the term *severe thunderstorm* or *severe local storms,* but only the chief of the U.S. Weather Bureau, or, in his absence, the chief of the Forecast Service could use the phrase "conditions are favorable for destructive local storms." The restrictions remained in place until 1938.

A tornado can be seen in the swirling column of dust and debris under this supercell mesocyclone near Akron, Colorado, on June 14, 1990. Looking a bit like an alien spacecraft beaming up samples of earth, this huge storm spawned at least five tornadoes, as well as hail the size of baseballs. The rotating wall cloud and the updraft column (above the funnel) can be seen clearly. Photo: Eugene McCaul Jr.

During this period, citizens of tornado-prone areas learned to rely on their senses, observations of nature, and *The Farmer's Almanac* for weather predictions. Folk wisdom had taught Plains residents that the sky would often turn green, the wind would cease blowing, and animals would become agitated just before a tornado. One Apache method of weather forecasting was reading patterns in bear grease. According to this theory, animal cells respond to the weather even after the animal has died. The bear grease in animal bladders or jars formed various patterns on the sides of the containers, and reportedly tornadoes would follow observations of funnel shapes that appeared in the grease.

The U.S. Navy Department of Aeronautics issued a directive on tornadoes in March 1943. The report acknowledged that tornadoes were impossible to forecast because of their highly localized nature but that if naval station personnel had adequate warning of an approaching tornado, they could help pilots fly their aircraft to safety. In light of this directive, the navy's aerological officer at Hensley Field in Dallas, Texas, requested a network around the naval installations in the Dallas-Fort Worth area; the Weather Bureau complied in August 1943. The following spring, Civil Defense officials met with their counterparts from the U.S. Weather Bureau, the Army, and Navy to discuss Civil Defense's participation in the Severe Storm Warning Service. At the meeting, Weather Bureau personnel agreed to train and coordinate the observers that Civil Defense would provide, and the U.S. Army Air Corps and Navy advised the Weather Bureau where they needed networks. By February 1945, some 162 Severe Storm Warning Service networks with 3,685 observers were operating.

Methods of spreading the warning varied during the 1950s from company and fire department sirens to rail-yard whistles. Radio, however, was the most common method of disseminating tornado warnings. Broadcasting of weather reports over commercial radio stations began in the early 1920s. Only 20 of the nation's 36 commercial stations had a license to carry Weather Bureau forecasts by 1922, but the ultimate plan was for at least one station in each state to distribute official weather forecasts and warnings. The bureau certified all 140 existing stations for weather broadcasts in January 1923. Over the next few years the bureau and radio stations created links that enabled stations to cover local threats in a timely manner.

Most storm-spotter networks phoned radio stations in their areas to report tornadoes, and the radio stations relayed the warnings to their listening audiences. Over 95 percent of American households had radios by 1950. Only 5 percent of American households had a television in 1950, although both the number of sets and stations increased dramatically during the decade. This new medium, which would become the public's chief source of severe-storm warnings in the future, cautiously entered its new role in Oklahoma City in 1952, when the chief of the Weather Bureau granted WKY (now KTVY), the city's sole television station, permission to use Tinker Field's tornado predictions on the air. To avoid a panic, the WKY announcer would preface all warnings or weather information with the following statement: "We will pause a moment in our regular schedule in order to bring you some important weather information."

When a hook echo appeared on the Weather Bureau's radar screen in Wichita Falls, Texas, on April 3, 1964, station KAUZ's cameras began a live scan of the skies. As a spectacular twister roared through parts of the city, the station rewarded its viewers with the first live television broadcast of a tornado.

—*Marlene Bradford*

This enormous tornado was photographed in July 2002 from a single-engine plane during a cloud-seeding project. The tornado stayed in a rural area 15 miles northwest of Colby, Kansas, and damage was limited to a wrecked irrigation system and uprooted trees. Photo: Jason LaFontaine/Painet, Inc.

ROBERT FITZROY & THE SCIENCE OF WEATHER FORECASTING

LEFT: *Wild seas of the Drake Passage off the southernmost tip of South America. When Robert Fitzroy was in command of the HMS* Beagle, *he surveyed the coast of South America and produced charts so good that they were still in use during World War II.*
Photo: Galen Rowell/Mountain Light

ABOVE: *Painting of Robert Fitzroy, inventor of the weather forecast.*
Photo: Courtesy of the British Crown, © 2007, the Met Office

In 1836, Charles Darwin had been at sea for five years, traveling aboard HMS *Beagle* with a man who desperately wished his friendship. In fact, he had invited Darwin along as ship's naturalist precisely so that he could have someone to talk with, since the ordinary sailors of the British Navy could hardly be expected to keep up their end of a conversation. But Darwin and his benefactor had to struggle, too, to find things to talk about over those five long years, for Darwin, impatient and preoccupied, was well on his way to declaring his iconoclastic theory of evolution, while Robert Fitzroy, *Beagle*'s captain, was a religious fundamentalist who would sooner put the young scientist on an ice floe than admit that the Bible might be anything other than the literal truth.

That is how history remembers Fitzroy, if it remembers him at all. The image is quite unfair, and quite misleading. Darwin, a clergyman's son, was as religiously inclined as Fitzroy, while Fitzroy was well versed in science and expressed his own doubts whether earth had really been created, as the fundamentalists had it, in 4004 BC. Their positions solidified in opposite corners only long after they had returned to England. Then, in 1860, Fitzroy, by now an admiral in the Royal Navy and an ardent anti-evolutionist, turned up at a lecture given by Darwin at Oxford University and turned it into a grand debate—one that Fitzroy eventually lost.

Born in 1805, Fitzroy was well educated and interested in the world around him. He earned a perfect score on the exam that allowed him to graduate from naval college, and, when given his choice of assignments, he asked to join a survey of Tierra del Fuego, the little-known land at the southern tip of South America. There Fitzroy experienced, among other phenomena, the howling wind known locally as the *pampero*, which he would later describe in exacting scientific detail. His reports from Tierra del Fuego were so thorough and well written that he was called to the Admiralty before the Royal Navy's hydrographer, a man named Francis Beaufort.

Beaufort presented him with an assignment: Fitzroy was to return to South America on HMS *Beagle*, taking wind measurements along the coast, using a scale that Beaufort had devised—the Beaufort scale.

The *Beagle* was to be outfitted with a newfangled invention known as a lightning rod, as well as theodolites, barometers, and chronometers, all to keep an eye on the weather. When he was finished with South America, Beaufort instructed, Fitzroy was to do the same sort of surveying in the South Pacific, along the African coast, and indeed anywhere that English sailors might venture—and to take his time about it.

Thus began the five-year voyage that Charles Darwin made famous. On this and other scientific journeys, Fitzroy carried out his duties in exemplary fashion, so much so that in 1854 he was appointed to head a new department of the government's Board of Trade: an office devoted to meteorological statistics and research. Fitzroy immediately set about placing weather-observing equipment on ships in every port in England, eventually recruiting some 80 captains to make standardized measurements of the weather wherever their voyages took them. While he waited for the first reports to come in, he devised a new method of recording weather data, which he called "wind stars," showing wind directions and intensities around the world. Ships' navigators came to swear by his careful charts, which saved countless lives as they warned sailors of rough passages.

Fitzroy went on to invent a barometer now called the Fitzroy storm glass, still an accurate means of foretelling heavy weather. Of more lasting importance, he wedded meteorological science to advances in communications technology, devising the first scientific weather forecast. His method used the telegraph to transmit data from far-flung weather stations, which in itself was not an original idea; Fitzroy's contribution came in plotting that data in relation to the prevailing wind, demonstrating that strong weather systems generally moved over the British Isles from the west and moved at a pace that could be calculated. Thus a storm bursting over Galway Bay at Sunday breakfast, say, could be predicted within a reasonable margin of error to arrive in London by tea time. Once again, the ability to warn ships of impending storms and their approximate time of arrival saved many lives.

Fitzroy gathered the fruits of his long years of observation and experimentation in *The Weather Book*, published in 1862, in which he coined the phrase "an ocean of air." Almanac writers used his charts to devise weather forecasts that were as accurate as could be, given the time and technology; soon the *Times* of London would publish weather forecasts daily, even if its editors were in the habit of blaming Fitzroy personally when the forecasts were off.

Recognized as one of the great pioneers in scientific meteorology today, Robert Fitzroy received little praise for his work in his own time. Melancholic, in poor health, and on the losing side of the Darwinian debate, he took his own life on April 30, 1865, at the age of 59. He was forgotten for years, but then happily rediscovered. Today the British government's Meteorological Office and National Meteorological Library and Archive are located on Fitzroy Road in Exeter, England, and modern atmospheric scientists proudly acknowledge him as the founder of their discipline.

—*Gregory McNamee*

The Coast Guard Cutter Tamaroa*'s rigid hull inflatable rescue boat is sent to help the sailing vessel* Satori*, which needed help about 75 miles south of Nantucket Island after being caught in a northeaster-like storm that raked New England on Halloween week. Photo: Weatherstock*

Stories the trees tell can sometimes rewrite history. By studying the rings of trees, it's possible to reclaim some of earth's oldest climate records.

READING BETWEEN THE LINES

LEFT and ABOVE: *This bristlecone pine lies alone on the side of a hill high in the White Mountains of California. It is a fine example of* krummholz*—a term mountain geographers use for trees formed by the wind (the German translation for krummholz is "bent wood"). Locked within the ancient bristlecones' rings are records of past regional droughts and temperature swings. Photos: Ed Darack*

It twists upward from the rocky face of Ontario's Niagara escarpment, more bare trunk than tree now, and yet each spring, leaves still flourish from its few still-living branches. The process is remarkable, since this ancient northern white cedar (*Thuja occidentalis*) first sprouted from the cliff in AD 998. During the intervening ten centuries, this aging tree has recorded every rainfall, blizzard, drought, and wildfire that swept this region—a priceless meteorological record. It's all in the tree rings.

The same kind of weather data is exposed in the cut ends of the branches you prune from that tree in your yard, or in the stump of a Christmas tree. The grain in wood is simply these annual rings, cut obliquely; the figured wood of an old cabinet or desk really shows us swirling patterns of weather past.

Dendroclimatologists are the experts who can tell us this weather. Repositories such as the International Tree-Ring Data Bank are the official "library" for their findings, maintaining data collected by thousands of experts who have studied virtually every species of tree across the planet. It is, in effect, a forest of knowledge. And in every sample, variations in the tree's grain preserve a history of weather, often in startling detail.

For example, the closely spaced rings in a Rocky Mountain juniper (*Juniperus scopulorum*) recall a long-ago dry spell. The dark inclusions partly obliterate the lines. These are burn scars. This forest so wanted for rain that it caught fire. Based on the annual rings (each ring is composed of early and late season growth), we can pinpoint the blaze to July or August 1779. Many dendrological samples are centuries old, the last living witnesses to past events. Think of a tree as a kind of weather chart recorder that records very slowly.

The basic theory sounds simple: wide rings in a tree indicate wet seasons, narrower ones mark the dry years. Yet the complexities of a forest (each tree lives in its own microenvironment, and even woodpeckers can change how a tree grows) make turning tree rings back into weather more like trying to glue autumn leaves back on the trees where they fell. Raw measurements of rings must be taken from many trees in an area, then reduced to graphs, called skeleton plots. Measurements can be taken from living trees (which are cored using tiny probes, to avoid damaging them), as well as from dead trees, which stopped recording weather long ago.

Then similar, repeating patterns among these samples are correlated in order to "synchronize" the trees. This is how years are assigned to rings. It's something like breaking an enemy code using many tiny messages. The advantage lies in stringing the histories of different trees together, producing a plot of weather spanning much longer than the oldest living trees. These are called chronologies. It is all the brainchild of A. E. Douglass, an astronomer who developed the theory that tree rings were records of solar cycles of growth. He eventually determined that by studying the rings, one could reclaim some of earth's oldest climate records.

The stories the trees can tell sometimes rewrite our history. In rings corresponding to the year 1588, from a stand of bald cypress (*Taxodium distichum*) in Virginia's tidewater region, David Stahle of the University of Arkansas Tree Ring Lab has discovered evidence of the worst drought to strike the Americas during the past eight centuries. With this finding, the disappearance of the fabled "lost colony of Roanoke" may finally be solved, for the doomed colonists vanished from that spot in that very year.

Far more ancient timbers, preserved where they fell, can reconstruct weather from the prehistoric past. Stahle is currently exploring the remains of a 50,000-year-old forest, buried beneath modern Maryland. Once chronicled, the rain and heat that drove Pleistocene herds across ancient America will be charted nearly as accurately as last month's average rainfalls. How these results may transform our familiar "museum" images of the Ice Age and early humankind, only time and the trees can tell.

Still other findings prove as current as the nightly news. Each spring, wildfires are reported raging across the western states. Naturally, we blame the droughts. But is the weather changing, or have we changed the forests by logging too much old growth, leaving the forest less able to bear these fires naturally? At the University of Arizona Laboratory of Tree-Ring Research, Paul R. Sheppard and his team are exploring the relationship between rainfall and fire records in old Western forests to understand fire and weather in the preindustrial age better. If fire has changed but weather has not, we may need to reexamine our forest management policies.

Like discovering an old diary stuffed in the rafters, an old house or barn inadvertently preserves a record of the weather. It's stored in the grain of the beams and siding that make up the structure. At the Laboratory for Tree Ring Science at the University of Tennessee, Henri D. Grissino-Mayer has analyzed timbers from an aging frontier dwelling, long purported to be the log cabin where Abraham Lincoln was born. Is the verdict in the rings? The 16th president may have been "honest" Abe, but this cabin wasn't. Its timbers were felled in the 1840s; Lincoln was born in 1809. Perhaps this later cabin contains an occasional authentic log—making it, in Lincoln's own words, "a house divided against itself." It would seem that the history of weather, as stored in the trees, still has much to tell us about our past, including an occasional breath of fresh air.

—Nick D'Alto

Ponderosa pines rise into the Sierra sky near Sonora Pass, California. Photo: Ed Darack

THE INVENTION OF DENDROCHRONOLOGY

By training, Andrew Ellicott Douglass (1867–1962) was an astronomer. In 1894 he was hired by Percival Lowell to select a location and to oversee the construction of a new observatory—the great Lowell Observatory in Flagstaff, Arizona. There Douglass spent seven years assisting in the discovery of distant stars and planets, until an unfortunate dispute with Lowell led to his dismissal.

Resourceful and well rounded, Douglass made ends meet by teaching Spanish and geography at what is now Northern Arizona University while serving as a probate judge in Flagstaff. Looking for some way to connect the record of sunspot cycles to climate, it was among the tall ponderosa pines of campus and city where he developed the theory that tree rings were records of solar cycles of growth: the wide rings of certain species of trees recorded wet years, narrow rings dry years. Douglass found that he was able to correlate tree ring patterns with the time of various historical events, both natural and anthropogenic. In all events, each year a tree adds a new layer of wood, a ring that can be counted. From this observation, Douglass conjectured that by studying the climatological record of a region such as the Colorado Plateau, predictions could be made about future weather cycles.

Douglass moved south to Tucson in 1906 to teach at the University of Arizona. There he elaborated the science of "dendrochronology: which uses tree rings to determine the age of a particular piece of wood." He developed a ponderosa tree ring chronology dating back hundreds of years, which enabled archaeologists to use timber taken from sites such as Mesa Verde to date their construction—an important window on prehistory.

Tree ring analysis can also be used to diagnose the effects of air and water pollution and to adjust irrigation cycles in arid-lands farming. At present, the chronology—established mostly from ponderosa and bristlecone pines—has been extended to 7000 BC, and efforts are being made to push that date back even further, to the ending of the last Ice Age.

—*Nick D'Alto*

With the right data, predicting waves isn't so hard. The hard part is forecasting local wind and weather conditions which make the difference between ridable waves and storm surf.

PREDICTING WAVES

In the last decade, both the science and the art of surf forecasting have become incredibly precise for surf spots from up and down the coast of Malibu in California to Cape Hatteras on the East Coast to Hawaii's pipeline and spots in Australia, Fiji, South Africa, and Europe. Culling data from automated buoy reports, ship reports, wind information from satellites, and the bathymetry of specific surf spots, forecasters have created computer models that can predict the tide and quality of waves days and weeks ahead of time. Tens of thousands of surfers throughout the United States are plugged into Internet swell forecasts that have made finding the perfect wave as simple as typing on your computer.

"We're getting to the point that we are now able to predict to the hour when a large swell is going to hit," says Sean Collins, a veteran Huntington Beach forecaster and corporate officer with Surfline (www.surfline.com), an online weather and wave forecasting business. The company has eight forecasters on duty with the sole job of predicting waves. In the late 1970s, Collins got an underground reputation among surfers for his uncanny "spot-on" predictions of southern swells on California beaches generated by storms off the coast of New Zealand and Antarctica. Because the waves are generated thousands of miles away, it is often difficult to pinpoint where they will hit. Exact measurements of the strength and duration of those storms is vital. Underestimating the actual wind speed in a storm brewing at sea by just five miles per hour could result in a significant error in estimating the strength of waves.

Waves are a visible display of the energy in bodies of water. In the ocean, waves represent about one-third of all the energy available, with the other two-thirds linked to currents. Scientists are still investigating the many forces that cause and influence waves. Wind is responsible for most typical waves, first creating small ripples on the surface of the water, minor variations that are small enough to be flattened by surface tension unless they can grow in size. Their growth varies, depending on the temperature and salinity of the water. Ripples are usually not considered waves until the wavelength—the distance from crest to crest—is at least four inches, about the width of the average person's hand.

Crashing coastal waves like these at Pt. Pinos, Asilomar State Beach, California, are becoming easier to predict for veteran forecasters. Photo: Rich Reid

CLASSES OF WAVES

Type 1
Surface Waves include all normal surface movement caused by the wind, including ripples, swells, and waves.

Type 2
Internal Waves are created by the movement of subsurface currents over irregularities on the sea floor.

Type 3
Seismic Waves can be triggered by earthquakes or underwater landslides (also called "tsunamis").

Type 4
Solitary Waves are usually stable, single waves resulting from some human activity or from a tidal surge moving over an underground obstacle.

Type 5
Tide Waves are affected by the cycles of the moon and the sun.

Type 6
Planetary Waves are influenced by the effect of the earth's spin on ocean currents.

RIGHT: *Massive waves hit a pier at Atlantic Beach, North Carolina, as Hurricane Isabel approaches. Photo: Warren Faidley/Weatherstock*

FOLLOWING PAGE 52: *A violent sea storm caused by Hurricane Isabel rips into the Atlantic Ocean. Photo: Warren Faidley/Weatherstock*

FOLLOWING PAGE 53: *Huge waves crash along the southern coast of California, north of Los Angeles during the El Niño of February 1998. Photo: Warren Faidley/Weatherstock*

Once a wave begins to form, its potential height is determined by three factors: the speed of the wind, the distance the wind can blow across the surface of the water without interference, and the length of time the wind blows without changing either direction or velocity. The ultimate size of a wave is limited by the very force of the wind that creates it; the tops of waves that become too steep are blown over, hence the term "whitecaps." Some waves eventually turn into swells, the correct term for waves that have moved outside the area where they were generated. Waves actively being produced by the wind vary considerably in size and shape, but as they settle into swells, the size and shape become more uniform. Even without the continual presence of wind, waves and swells gradually lose energy because of the drag created by the turbulence of the water in which they move, a process known as "wave dispersion." But large swells—those created by big storms—can travel long distances, literally from one end of the earth to the other, without fading away.

Collins's company has pioneered predicting monstrous surf, waves in excess of 40 or 50 feet high and as tall as 100 feet. When the waves are that big, surfers cannot even paddle into them and they must rely on Jet Skis to tow them into the fast moving walls of water. Finding storms that generate swells of that size is not difficult, but finding places where the waves will break undisturbed by the weather is another matter. As Collins states, only a handful of spots around the world can hold large surf. Five years ago, a reef 100 miles offshore from southern California was ridden for the first time because of the surf and weather forecasts made by Sean Collins and his team. "That was something I'm very proud of," he said.

—Scott Hadly

LIGHT

LIGHTNING, RAINBOWS, AURORAS, COMETS, AND RAYS

"Daddy, why is the sky red?" might query a young descendent of Earth residing in a future Martian colony. And why blue here on Earth, except when it's orange, red, pink, or gray? For those taking the time to look skyward, a decided minority in our frantic, increasingly urbanized society, the celestial dome displays a never-ending kaleidoscope of tints, hues and textures, all in High Definition, without monthly cable or satellite charges.

The light detected by our eyes is just a fraction of the electromagnetic radiation spectrum emitted by our nearest star. Visible light is absorbed, scattered, reflected, and refracted by molecules, water droplets, ice crystals, and atmospheric layers, which extract the spectrum's rainbow colors. Shorter than violet, UV wavelengths can burn our skin while those longer than red can be felt radiating from a brick wall releasing the heat from a day's intense sunshine. Highly energetic particles streaming from the solar nuclear furnace energize the upper atmosphere into multi-hued curtains draped around both poles, not only on the third rock from the sun, but on Jupiter and Saturn as well.

Once thought a harbinger of war, the aurora sky show gets two thumbs up from all critics, and is also a reminder there is indeed "weather" in near space. Cosmic radiation and magnetic disturbances can cripple satellites, expose astronauts and transpolar airline passengers to worrisome radiation dosages, and trigger massive power outages in the intertwined electrical grids on the planet below. The same solar wind which bombards our upper atmosphere vectors the glowing tails of comets like Hale-Bopp to always point away from the sun which they orbit.

LEFT: *The vertical streak of reddish light seen here in Norman, Oklahoma, is called a sun pillar. This relatively rare phenomenon is caused from reflected light on the tops and bottoms of vertically aligned, falling ice crystals. Photo: Charles A. Doswell III*

FOLLOWING PAGES: *A midsummer storm coming south up the valley to the Yampa Valley in northwest Colorado creates a magnificent rainbow over the hayfields. The Yampa Valley and surrounding area contain several geothermal hot springs. Photo: Rod Hanna*

Once an omen of pestilence, Comet Halley's return as predicted by Sir Edmond is now celebrated; we, meanwhile, send probes to sample icy, rocky cometary surfaces and worry if after 65 million years, another really, really big one might impact us. Everyone's favorite, the rainbow, once sent a Franciscan monk to prison for life. Around 1265, Roger Bacon displeased the ecclesiastical authorities by daring to demonstrate how raindrops refracted sunlight to decorate the heavens without need of direct Divine intervention.

Yet one light in the sky, brighter and hotter than the sun, should be feared, or at least respected. We now know neither Thor nor Jupiter nor Zeus hurl the lightning flashes which strike the U.S. 30 million times yearly, killing perhaps a hundred Americans and thousands worldwide, with damage mounting into the billions. Following the 30:30 rule can save you. If the time delay between the brilliant flash and the resounding boom of the thunder is less than 30 seconds, the next bolt could well strike where you are standing. Seek safety inside a well-constructed building and avoid conducting objects with pathways to the outside. Stay inside until 30 minutes after the last rumble of thunder. Many fatalities occur as people venture outside to admire the receding storm, only to encounter one of the many strikes occurring outside the storm's rain footprint.

Perhaps that thunderstorm was triggered by a sea breeze on a tropical coastline, where evening finds you relaxing on the beach as the sun sets slowly into the ocean, allowing a glimpse of the elusive green flash. This brief emerald apparition results from sunlight's passage through layers of moisture and temperature near the sea surface. Or maybe it is just Mother Nature giving you a knowing wink because you took time to slow down, smell any available roses, and especially watch the never ending light show that is our sky.

—Walter Lyons, CCM

NATURE'S FIREWORKS

Of all the dangers of a thunderstorm, lightning can be the greatest. There are about 1,800 thunderstorms in progress over the earth's surface at any given time, and lightning strikes about 100 times each second. Some scientists believe that this constant bombardment of lightning is critical in maintaining the electrical balance of our atmosphere. The majority of deaths occur when people fail to seek shelter in open areas such as golf courses or when people seek shelter under isolated trees.

Roughly 100 people per year die in the United States from lightning strikes, with an average of 258 receiving lightning-caused injuries. Lightning also causes more than $200 million in property losses to structures, forests, and livestock annually.

Can there be lightning without thunder? It's impossible. A flash of lightning can be as hot as 50,000 degrees Fahrenheit. That intense heat causes the air around it to expand violently. The expanding air always causes the sound waves we call thunder. If you see lightning and don't hear thunder, it means that there was a rumble that was too far away for you to hear. Lightning appears to flicker because the lightning stroke is usually composed of a series of rapid electrical exchanges between the cloud and the ground.

The third eruption of Galunggung, a stratovolcano located on the west side of the island of Java, Indonesia, provided this spectacular display of lightning on December 3, 1982. Lightning that occurs during the volcanic eruption is often due to friction amongst the dust particles ejected from a volcano. This eruption resulted in 68 deaths. The first recorded eruption of Galunggung in 1822 produced a 14-mile-long mudflow that killed 4,000 people. Photo: R. Hadian/U.S. Geographical Survey

Even though this seems to be the opposite of what we think we see, most lightning bolts travel from the ground up to a cloud. The jagged paths they follow, however, are formed from the cloud to the ground a split second before each bolt's discharge. Think of its jagged path as a channel. A lightning channel is formed in steps, each about 50 yards long, moving from the cloud toward the ground. Each of these steps jumps forward in less than a millionth of a second, and only about 50 millionths of a second passes between steps.

Eventually, the negatively charged channel gets close enough to the ground (or to something projecting from the ground, like a golf club) to attract the positive charge from the ground. The stroke of light, resulting from the negative and positive charges making contact, travels up the channel—at speeds of 20,000 to 60,000 miles per second.

One major misconception about lightning is that it does not hit the same place twice. The Empire State Building in New York City is hit about 23 times each year. Commercial aircraft are struck by lightning an average of once every 5,000 to 10,000 hours of flight. Although flying through a lightning storm can be scary, most modern aircraft are safe from lightning dangers.

If you ever feel the hair stand up on the back of your neck, move your position quickly. What you are feeling is the electricity of the channel through which the lightning will travel.

—*Warren Faidley*

LEFT: *A lightning storm and towering cumulonimbus clouds move across Canyonlands National Park, Utah. Photo: Richard Kaylin/Getty Images*

RIGHT: *Storm chaser Hank Baker took this photo on May 30, 2006, in southwest Oklahoma near the town of Granite from his storm-chase truck, which was only 20 yards away from the lightning strike. He had been following the large supercell storm that was heading due south in western Oklahoma. Photo: Hank Baker*

U.S. LIGHTNING EXTREMES

- **Longest lightning bolt observed:** 118 miles (near Dallas, Texas)
- **Highest frequency of bolts in a year:** Between Tampa and Orlando, Florida
- **Lowest frequency of bolts in a year:** Sections of the Pacific coast
- **Average number of cloud-to-ground strikes observed in a year:** 30 million
- **Longest fulgurite (glassy, solidified path left when a bolt hits the ground):** 17 feet (near Gainesville, Florida)
- **Maximum power of a bolt:** 300,000 amps, 200 million volts
- **Amount of heat created when lightning is generated:** Up to 60,000 degrees Fahrenheit.

LEFT: *Life-giving monsoon storms in the desert are a welcome and sometimes magical sight. This lightning storm is arriving over "The Mittens" in Monument Valley, Arizona. Photo: Dan Heller*

RIGHT: *A fast-moving April squall whips up from Mexico to Organ Pipe Cactus National Monument in southern Arizona. The high winds even managed to move the Organ Pipe cacti arms as the rosy light engulfed this section of the Sonoran Desert. About half of this desert's meager eight inches of annual rainfall comes in these brief but violent thunderstorms. Photo: Jack Dykinga*

FOLLOWING PAGE: *Lightning of this type, "Anvil Crawler," can splay its raw beauty over several hundred miles of sky with each discharge. This brooding anvil cloud and lightning display was part of a line of activity near Norman, Oklahoma, in which baseball-sized hail was produced. Photo: Gene Rhoden*

ABOVE: *Six displays of lightning in the desert (top row) and in southwest Oklahoma (bottom row). Top Row Photos: Heather Gravning; Bottom Row Photos: David Ewolt*

Ancient Africans believed that people who were struck by lightning had incurred the wrath of gods. These lightning strikes were considered bolts of justice.

LIGHTNING CURIOSITIES

Lightning is a weather occurrence that has a long association with humans. The ancients believed that lightning was a power of the gods—but usually only the most powerful of those gods, such as Zeus or Thor. Those who were born during lightning storms, such as Augustus Caesar, were regarded as having the special attention and favor of the gods. Today, we understand lightning as a natural physical phenomenon—a redistribution of electrical charges between cloud and ground or cloud and cloud. Yet, even with our improved understanding of this natural "electric circuit," there are plenty of oddities and curiosities involving lightning that defy easy explanation.

Some of the powers of lightning have been known even since ancient times. For example, lightning tends to hit higher objects (because positive charges are drawn to highest points) and water-soaked objects (because electrical charges flow easier through water than air). We have evidence that many civilizations have understood these basic principles, although perhaps not the physics behind them. For instance, one of Genghis Khan's laws to his people forbade them to bathe or wash garments in running water during a thunderstorm. This is very likely advice directly linked to the lightning dangers associated with being around water in a thunderstorm. Similarly, the Greek historian Herodotus wrote around 450 BC that "you may have observed how the thunderbolt chastises the insolence of the more enormous animals whilst it passes over without injury the weak and the insignificant." This again shows ancient knowledge of lightning's dangers, noting the effect size (as in height) plays in lightning strikes.

Lightning has also had a presence in legends, perhaps exaggerating real experiences. In one historical account, for example, the great Roman emperor Marcus Aurelius achieved an impressive lightning-based victory against the Barbarians in AD 174 when it first appeared that nature was working against him. Marcus and his army were in a very distressing position, surrounded and completely besieged. Given that the Barbarians were on horse and thus more maneuverable than the Roman infantry and that the summer's heat and dryness were slowly killing his army, Marcus's fortunes were not bright. So he called one of his advisors, a great Egyptian wizard, to craft some powerful weather magic. The mystic sorcerer conjured a huge rainstorm, which formed over the Roman ranks.

It's "raining" lightning over Tucson, Arizona, as the summertime instability in the air, well known in the American Southwest, helps nature put on one of her most awe-inspiring light shows. Photo: Thomas Wiewandt/Getty Images

Initially, as the rain poured down, the Romans turned their faces upward to drink. Some held out their shields and helmets to catch the life-giving rainwater. The Barbarians realized that this was the perfect time to attack and they charged the preoccupied Romans. As they raced their horses towards the Romans, a terrible massacre appeared to be in the offing. Again, Marcus ordered his Egyptian wizard to do something. Calmly, the wizard morphed the rainstorm into a lightning-filled thunderstorm. The Barbarians, mounted on horses, were taller than the Roman infantry and the lightning seemed to preferentially select the attackers. Believing that the gods were with the Romans, they lost heart. The tables turned, the Romans set upon them with a vengeance, and Marcus Aurelius won the battle. As the story goes, he renamed that army the "Thundering Legion" in recognition of its link to the miraculous thunderstorm.

Bizarre, even unexplainable, occurrences have also been linked to lightning. Camille Flammarion, a noted astronomer of the 19th century, collected a huge number of lightning anecdotes. He reported that at Altenbourg, Saxony, in July 1713, lightning struck a woman who was pregnant. She delivered a child some hours afterward who, according to reports, "was half burnt, and whose body was all black." The mother recovered after the traumatic event. In a similar fashion, there are a number of stories about people seemingly "petrified" in place after being struck by lightning.

The primary cause of death by lightning isn't burns but cardiac arrest. In 1988, for example, a 14-year-old girl was riding her horse in a competition in a small town in Denmark when she and her horse were struck by lightning. Reportedly, the horse was killed instantly, the rider suffered heart failure, and several spectators were knocked down. After receiving CPR, however, the girl recovered with only burns on her chest and stomach.

Such burns in earlier times were thought to be "lightning photography." According to this belief, the force of the lightning could imprint an image of the surrounding area onto the skin of the unfortunate victim. Andre Poey, a noted 19th-century hurricane expert and collector of weird lightning stories, recounted that in September 1857, a French peasant girl who was tending a cow took refuge under a tree during a severe thunderstorm. Lightning struck the tree, the girl, and the cow. The cow was killed but the girl recovered. Reportedly, however, the "distinct image of the cow" was observed upon the girl's breast after the incident. Today we know that such marks are likely a form of electrical bruising and are not related either to the victim's surroundings or to the victim's internal arteries or veins.

Lightning is a strange phenomenon that continues to fascinate us. We continually learn new and exciting aspects of lightning, such as recently identifying the types called sprites, blue jets, and elves, with ever-expanding scientific understanding of how they are formed. Yet oddities still happen. As the Associated Press reported several years ago, a Norwegian couple was in bed during a severe thunderstorm when a lightning bolt struck their bedroom. According to the couple, the room lit up "like 10 welders' torches" with smoke from charred electrical sockets drifting throughout the room. They were not injured. Apparently the mattress and pillows somehow insulated them from receiving an electrical shock through the iron bed frame, but they did elect to spend the rest of the night on their sofa in another room. Just another reminder that in spite of the fact that we know more, lightning still has the power to surprise us.

—Randy Cerveny

Broiling storm clouds covering the eastern half of the sky in Nice, France, are illuminated by a brilliant full moon. Photo: Dan Heller

BASIC LIGHTNING SAFETY TIPS

Recommended by the National Weather Service:

1. Seek shelter immediately if a thunderstorm (or storm clouds) threatens. Golf courses, playing fields, hiking trails or any open area are especially dangerous.
2. If caught in the open, avoid tall, isolated trees. Crouch down and keep a low profile. Groups of people should keep apart. Avoid hilltops.
3. Get out of water and off small boats.
4. Avoid metal objects such as golf clubs, tractors, fences, metal pipes, tools and clotheslines.
5. Stay away from open doors, windows, radiators, bathtubs/showers, and sinks. Do not use plug-in electrical appliances.
6. Avoid using the telephone, except for emergencies. Lightning can travel through phone lines.
7. If you feel your hair stand on end or your skin tingle, lightning may be about to strike. Crouch down and cover your head. Do not lie flat.
8. Learn CPR. Many people who have been hit by lightning can be revived by basic CPR. People who are hit by lightning do not carry an electrical charge.

LEFT: *A tropical storm that had pounded Baja, California, drifted north toward California's Central Valley, igniting this rare lightning storm over San Francisco on September 8, 1999. Long-time residents claim that it was the largest display of lightning they had ever seen. Photo: Douglas Keister*

RIGHT: *Just like a vintage movie, this image captures the romance of the landscape that lured settlers west. Shot in Arizona northwest of Tucson, this image captures a spectacular July lightning storm looking west toward the sunset and the Tucson Mountains. Photo: Keith Kent/Peter Arnold, Inc.*

Until recently, auroras were a great mystery and the source of much folklore and mythology. In medieval Europe they were thought to be reflections of the shields of warriors honored in heaven.

EARTH'S MAGNETIC LIGHTS

The aurora borealis, or the northern lights, is among nature's most awe-inspiring sights. Equally impressive is the aurora australis, or the southern lights, visible in the Southern Hemisphere. Both are earthly phenomena, occurring high in the atmosphere, but are spawned some 93 million miles away in the sun.

Energetic electrically charged particles (mostly electrons, protons, and alpha particles, or helium nuclei) continuously speed outward from the sun at about a million miles per hour. This is called the solar wind. When there is particularly strong activity on the sun, typically corresponding to the sun's 11-year sunspot cycle, the solar wind will be particularly strong.

Since they are electrically charged, the particles (mostly energetic protons) that cause the auroras are attracted to earth's north and south magnetic poles. That is why the auroras are more frequently seen the closer you are to the poles. As these energetic particles spiral down along the magnetic lines of force, they interact with atoms and molecules in the upper atmosphere (mostly oxygen and nitrogen), causing them to glow or fluoresce.

A brilliant pallet of "aurora" colors light up the sky above the Talkeetna Mountains near Broad Pass, Alaska, in the early morning hours. The aurora borealis (called the aurora australis in the Southern Hemisphere) is a geomagnetic storm of electrically charged particles from the sun spiraling toward earth's magnetic north and south poles. A reaction between these earth-bound particles and earth's upper atmosphere causes them to fluoresce—to emit electromagnetic radiation, which is seen as visible light. Photo: Fred Hirschmann

Untold numbers of these fluorescing particles cause the aurora, which may appear as anything from a faint red glow to an elaborate fluttering curtain or flower-like pattern.

Each type of display can be classified into one of four phases. Phase 1 is a featureless glow. Phase 2 is a well-defined arc of light. Phase 3 is a moving curtain of light, possibly with streamers or rays that fade in and out. Phase 4 is a stage of irregularity in which the display appears broken, patchy, and flaming. Phase 3, the most spectacular and best to photograph, happens more often during years of peak sunspots. This phase can dramatically illuminate the entire sky and landscape manyfold in less than a minute, and to various degrees with colors. Unfortunately, this phase typically is the shortest in duration most of the time. Recently discovered large-scale winds at great altitude may be responsible in part for the various auroral forms.

The northern lights come in many colors, although greenish-white is the most common, followed by yellow and rarer red during intense surges.

Photographs show the colors more vividly than they actually appear, but they still cannot come near to showing the graceful motions of the auroral curtain and that of its details. These and the awesome spiraling and converging of the dynamic auroral plasma into a wondrous beautiful corona overhead are sights that not even Scheherazade of *Arabian Nights' Entertainment* could imagine.

—Jay Brausch

PHASE 1

PHASE 2

The aurora has four basic phases of behavior, which can come in any order in a given display. No display is exactly like another, and a display doesn't have to go through all four phases in a night. They are the glow (phase 1), the homogenous arc *(phase 2), the* rayed arc *(phase 3), and the* "chaotic" patches *(phase 4). Often, a display will appear to have a combination of phases, but for the sake of simplicity, the phase that is most dominant is what is recorded.*

Phases 3 and 4 are the active phases, and phases 1 and 2 are the quiescent (pre-surge) phases of behavior. The consistent recording and measure of the frequency of visible auroras from a given location over the long-term is also a measure of solar activity. The hyperactive sun by means of presenting unusually high amounts of auroras here is the greater indicator of the global warming phenomenon. Photos: Jay Brausch

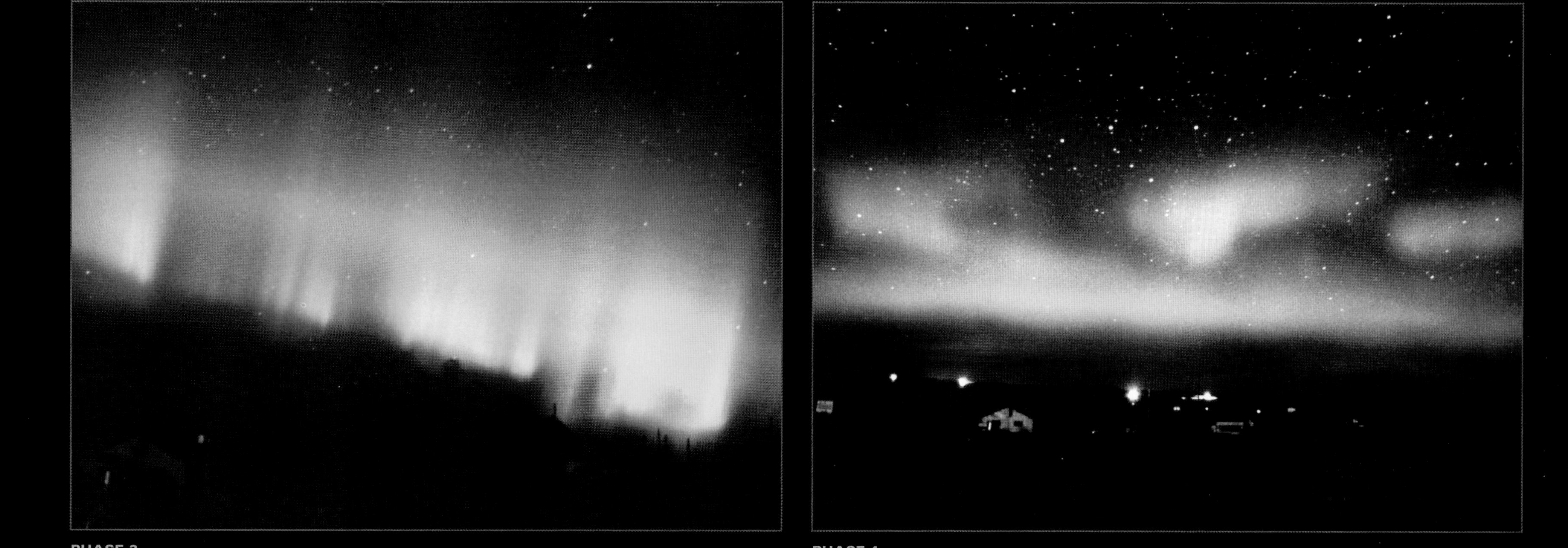

PHASE 3

PHASE 4

LEFT: *The aurora borealis, or northern lights, over the Barrens, Northwest Territories, Canada. Photo: Galen Rowell/Mountain Light*

ABOVE: *Auroras have evoked awe, fear, and wonder for untold millennia, but the true extent of their magnificence has been revealed in recent years through space exploration. The crew of Space Shuttle Discovery captured this view of the aurora australis, or southern lights, over the south polar region. This image from space shows the entire aurora oval. On Earth, only a partial view is possible. Photo: NASA*

LEFT: *In 1991, the Hekla volcano in Iceland erupted, sending ash to a height of 37,730 feet to meet auroras that were visible only about 62 miles overhead. The electrical discharges are visible in the night sky in a pattern that is roughly semicircular around the pole, and are often accompanied by strong geomagnetic disturbances. The occurrence of auroras follows a cycle similar to the sunspot cycle. Great auroral displays in mid-latitudes are usually associated with intense flares in the vicinity of sunspots. Photo: Sigurdur Stefnisson*

ABOVE: *Alaskan northern lights light up the sky and put on a beautiful display of color. Photos: Fred Hirschmann*

CAN YOU CATCH THE "GREEN FLASH"?

Have you ever seen a green star? Chances are, you haven't. But the light from all normal stars, including our sun, contains green light. In fact, most stars emit most colors of light. In the case of our sun, and stars like it, the combination of different wavelengths results in white or yellowish-white light.

But what about that green light? How do we know the sun has green light? Everyone has seen a rainbow with its arc of spectral co1ors—red, orange, yellow, green, blue, indigo, and violet. So when sunlight is bent by millions of rain droplets, through a combination of refraction and reflection, it is broken down into different colors, including green.

There is another way to see the sun's elusive green light, by observing a transient if not rare phenomenon called the "Green Flash." It sounds like a comic book superhero, and for most people it is just about as real. But this and other phenomena are real. Under the right conditions, the Green Flash may be visible at the top of the setting sun, seconds after the body of the sun has slipped below the visible horizon. (It is sometimes visible with the rising sun, too, but those observations are more difficult.)

When a rainbow breaks down light into its constituent colors, it does so with water droplets, which act as tiny prisms. But the earth's atmosphere itself can bend light, sometimes causing us to see different colors at different times. Look at a very bright star near the horizon on a warm, humid night. Chances are you will see that star flicker and change color very rapidly. The effect of the starlight passing through many different cells of atmosphere at slightly different temperatures and pressures causes different colors of light to be bent differently. In one split second the red colors may be bent in our direction, while in the next split second it may be the blue light. When the sun is near the horizon, its light is passing through thicker layers of atmosphere than when it is high overhead. Thus the effect of the air in bending sunlight is greatest at this time. So the sun's light can be dispersed into different colors. At this time you may also notice the sun appears flattened, which is another effect of the atmosphere.

PRECEDING PAGES 82–83: *A polar bear enjoys a stroll on a frigid November evening in Manitoba, Canada, as the illusion of three suns lights his way. This atmospheric phenomenon, called a "sun dog," occurs when ice crystals in the air bend sunlight at a 22 degree angle, not unlike a prism, and appear prominently when the sun is low in the sky. Photo: Steve Bloom*

LEFT: *This is the Green Segment which differs from the elusive Green Flash only in that the disk of the sun is still visible. In this striking image, shot from Mauna Kea Volcano, the sculpted shadows across the bottom of the sun are distant clouds over the Pacific. Photo: Kenneth D. Langford*

Why don't we see the sun change into all these different colors at sunset? The blue wavelengths are scattered by molecules of air. Because their wavelengths are comparable to the size of air molecules, they have a more difficult time finding a straight line. Instead, they are bounced off into all directions, which is why the sky is blue. On the other hand, orange and yellow light is not scattered but absorbed by the atmosphere (primarily by water vapor, oxygen, and ozone). So the setting sun typically doesn't appear orange or yellow, because the atmosphere absorbs those wavelengths.

The red light usually passes through, which is why the setting sun frequently appears red. But like red light, green light is neither absorbed nor dispersed. And when the red light is fading, for just a brief moment, the top of the sun may appear greenish. Sometimes this green edge of the sun appears to detach itself and float above the sun for a few moments. This is the "Green Flash." Although the Green Flash has been known for a long time, relatively few people have seen it because it requires a certain set of circumstances and is very fleeting. Few people are looking at the right time.

Can you see the Green Flash? Although the Green Flash is best seen through a telescope or binoculars, we emphatically do not advise this. It is never safe to look directly at the sun, either with just the eye or with optical aid. In the last few moments before setting, the sun's light is usually greatly attenuated, but anyone who looks at it must do so at his or her own risk.

The Green Flash and other color phenomena are best seen when there is a low, clear horizon without clouds, dust, haze, or any form of pollution. From a mountaintop or over the ocean are typically cited as the best observing conditions. The flash is a fleeting and somewhat capricious phenomenon, which some observers have tried for decades to see and have never succeeded.

—Larry Sessions

ABOVE: *Here the "Blue Flash" is seen over the Atlantic Ocean from Southampton, Bermuda. The blue color, perhaps with a hint of violet, is more likely to be seen when the air is haze and pollution-free. Photo: Kenneth D. Langford*

FOLLOWING PAGES 88–89: *This spectacular summer storm lit up the Great Sand Dunes of southern Colorado for some time, accompanied by noisy thunder but very little rain. Photo: E. R. Degginger*

HALE-BOPP: COMET OF THE CENTURY?

In 1997, Hale-Bopp graced the dawn sky of late February and earl March with its gas tail and a breathtaking increase in brightness. Then in mid-March, the comet crossed to the evening sky where millions more people saw it plainly with the unaided eye. Hale-Bopp posed magnificently with the moon in early April and remained visible withou optical aid until early May.

Schoolteachers described Hale-Bopp as the only thing they'd eve witnessed that got kids away from TV sets for a while. Hale-Bopp was the only modern comet bright enough to reach for many weeks into even the most light-polluted cities—a torch of hope from the world o Nature and grand perspectives.

Hale-Bopp's head—a cloud of glowing gas twice the sun's width in space—was one of the largest ever recorded. The head featured vast perfect "hoods" of golden sunlit dust that were visible even in fairly small telescopes for about two months. These were being puffed out by the icy nucleus of the comet. The nucleus produced about 100 times as much gas as Halley's Comet does at comparable distances from the sun and probably many more times as much dust.

Despite its tremendous true length, Hale-Bopp's tail appeared foreshortened by pointing mostly away from us. And the better scattering angles for the dust tails of Comet Bennett in 1970 and Comet West in 1976 helped make those tails bright for a greater apparent length than Hale-Bopp's. But in Hale-Bopp the brilliant section of tail was at least long enough and certainly bright enough to be plainly visible from big cities and to identify the object immediately as a comet.

Hale-Bopp never came close to earth or the sun. But its intrinsic brightness was rivaled by only a few others in the past 400 or 500 years. And Hale-Bopp was a "great comet"—a comet brighter than the Big Dipper stars when out of twilight in full darkness—for longer than any comet on record.

—Fred Schaa

LEFT: *The grandeur of the comet Hale-Bopp over the mountains. Photo: Paul Neiman*

WIND

TORNADOES, DUST STORMS, WATERSPOUTS, AND HURRICANES

The names of earth's winds are legion: cyclone, twister, tornado, willy willy, dust devil, waterspout, katabatic, blue norther, Santa Ana, foehn, sirocco, haboob, Tehuantepecer, chubasco, brubu, kona, laventer, bora, mistral, blad, williwaw, and even brickfielder. That last name given by the early residents of Sydney, Australia, for the gusty blows from the south bringing red, dust-filled air from the brick kilns outside of town.

Farmers, sailors, fisherpersons, or whoever spends time in the outdoors, learn their local winds well, with new zephyrs denominated with every change of geography. But all wind is basically the same: the movement of that fluid we call air from one place to another. Scientists can even quantify it mathematically rather easily on paper. Start with a pressure gradient force, take into account the Coriolis effect from our spinning planet, subtract the drag of friction and turbulence, and round it off by accounting for the centripetal and centrifugal forces of spinning vortices.

LEFT: *This classic F2 twister near Watonga, Oklahoma, is shown here at the beginning of its "roping" stage. The funnel was on the ground for 20 minutes, and was only 1 of 26 tornadoes in Oklahoma on that day. Photo: Hank Baker*

FOLLOWING PAGE: *A Kansas state trooper quickly throws his patrol car into reverse to avoid being sent to the Land of Oz by this menacing twister near Pretty Prairie, Kansas, on April 11, 2002. The tornado, which passed within 500 yards of the officer, had been on the ground for nearly 30 minutes when it suddenly changed directions. Photo: Jim Reed/Getty Images*

FOLLOWING PAGE 95: *This tornado was the first of 10 to touch down on May 29, 2004. It developed four miles east of Attica, Kansas, and moved north through rural farm country, lasting 24 minutes. The bottom half of the tornado was illuminated by the setting sun, creating a beautiful orange glow. The outbreak was caused by a lone supercell moving eastward. This tornado had an F1 rating with associated wind speeds of 80 to 110 miles per hour. Photo: Eric Nguyen*

Simple in theory, perhaps, yet the complexity of real-world air motion requires more than a little scientific ingenuity to measure and record the winds' every gust and ebb on all their myriad scales. We find the wind in thunderstorm updrafts suspending a giant hailstone, tortuously swirling around a home as a thunderstorm gust front tears across a suburb, advancing as a great army of air across continents with seasonal monsoons, roaring along as a jet-stream river girdling the globe, and as the gentle upliftings and subsidings of the general circulation which define where moisture-laden storms rampage or where clear, hot air endlessly sinks over barren deserts. Some winds of the tropics have personal names that will long be remembered by their victims: Mitch, Katrina, Iniki, and Tip. The thousand-plus tornadoes that annually assault our nation remain nameless, but their victims never forget the instant their lives were changed by the shrieking wind. By way of respite, dust devils and waterspouts seem almost playful, skittering above the scrub landscape of hot desert sands or the warm ocean.

Air may seem so tenuous, yet earth's winds do the heavy lifting of our climate system. Without the winds, the tropical regions would become unbearably hot and the poles numbingly cold. Winds carry moisture, making agriculture and civilization possible. Delay or failure of the Indian monsoon means drought and famine. Pacific winds create the coastal forests and snow packs of the American West. Sometimes winds hurling moisture against mountain barriers unleash torrential downpours, killer flash floods, and landslides.

The wind's potential to rip apart humanity's proud achievements increases with the square of its velocity as roofs are scattered across a neighborhood. Other less provocative winds transport great clouds of dust from central Asia, yellowing the western American skies each spring. The carbon black smoke of the Kuwaiti oil fires circumnavigated the world not once, but twice. Six inches of Georgia's topsoil are reputed to have blown in from the deserts of Africa.

Were it not for the semi-permanent bands such as the trade winds, ancient commerce and global exploration may have been constrained to landlocked migrations. Our world, our livelihoods, our everyday lives are perpetually influenced by the ceaseless wind—even during the calm before the storm.

—Walter Lyons, CCM

STORM DAMAGE

Mother Nature packs a wallop. The floods, wind, and hail that represent nature at its extreme are powerful forces that pose a serious threat to flora and fauna, the landscape, humans, and the structures in which they live and work. Modern technology provides significant protection against even the most extreme storms, but every year in the United States, storms still wreak havoc, producing billions of dollars of damage.

Hurricanes often get the most attention as devastating forces, but compared to thunderstorms, they are few and far between. The U.S. National Weather Service estimates that at any given time, about 2,000 thunderstorms take place throughout the world, but only severe thunderstorms are likely to cause serious damage. In the United States, about 10 percent of all thunderstorms are rated at this level, with winds of 58 miles per hour or more and/or with hail at least ¾ inch in diameter.

Tornadoes that accompany thunderstorms are well known for their dangerous wind, but more common are straight-line winds and microbursts, which can reach velocities of 100 miles per hour or more. About 40 percent of all property damage during thunderstorms is caused by this kind of wind, and annual losses from non-tornado winds are estimated to be about $200 million a year. By comparison, annual tornado damage is about $500 million a year. Most of the damage produced by wind occurs to the roofs and siding of buildings. A single shingle, lifted at its lower edge by wind gusts, exposes the underside of the overlying shingle; if the wind tears away the first, it creates a cascading chain of failure that can pull away dozens to hundreds more. With the strongest blows, the underlying sheathing or even the entire roof may be ripped away.

Despite the threat from wind, the greatest danger from thunderstorms is hail. Hail comes in various sizes, from pea-sized on up, and varies in density (number of hits per area) as well as impact speed (a factor related to how hard the wind is blowing the ice pellets when they hit). Although deaths from hail strikes are rare, damage to crops and structures is common. Annual crop losses in the United States alone are estimated to be at least $1 billion; non-crop damage averages about $350 million a year.

Most structural hail damage is to roofs. The major variables involved are the type of roofing material—asphalt shingles are the most vulnerable—and the angle of the roof. In general, the steeper the pitch, the

The record-tying nature of this tornado meant little to residents of Manchester, South Dakota, as their hometown disintegrated under the onslaught of this half-mile wide F4 tornado on June 24, 2003. Manchester, standing on the prairies since the 1880s, would never be rebuilt. Photo: Karen Leszke & Gene Rhoden

less the damage because falling hail produces less damage when hitting at an angle. These ice pellets create damage by knocking off the shingle's protective outer layer (usually a mineral coating), breaking off the edges, or penetrating completely through the material. In general, the warmer the temperature, the better asphalt shingles absorb hail impact without damage because they are more flexible when warm. Flexibility is also related to age: the older the shingles, the more likely they are dried out and brittle, increasing the potential for damage.

Crop damage from hail is linked not only to the size, density, and velocity of hail, but to the type of crop and the stage of plant growth. For a month or so after sprouting, for example, corn plants typically do not suffer much during a hailstorm because the plants are small and represent less of a target. From mid-summer on, however, it's a different story. Even a moderate hailstorm can ruin a corn crop because these plants do not recover when defoliated by falling ice. Soybeans, on the other hand, are better able to survive defoliation, at least early in a season. In order to assess the damage from hail, insurance agents who specialize in crop damage must use careful analysis that is specific to the crop that has been hit.

With or without tornadoes, a single severe thunderstorm can produce a tremendous trail of damage, and when multiple thunderstorms occur as outbreaks, the effects quickly amplify. The record for a single thunderstorm comes from a supercell on May 5, 1995, in the Dallas-Ft. Worth, Texas, area, with $1.1 to $2 billion in claims from a combination of wind, hail, and flooding.

Wind and hailstorms can spread destruction over many acres, accounting for their ability to rack up large losses. But lightning, although singular and narrowly focused, poses a more frequent threat. The most frequent range of loss is between $5,000 and $50,000 per strike (at least for those reported as causing damage); between 1959 and 1994, the National Weather Service lists 19,814 total damage-causing strikes. Most strikes hit only a single structure and cause limited damage per strike, especially if the structure is a residence. But up to 30 percent of all church fires are caused by lightning strikes, as are more than three-quarters of all accidents involving gas and oil storage tanks. When a single lightning strike produces the most expensive damage, it is usually because it triggers a wildfire. In the western United States alone, an estimated 10,000 such fires occur annually, with losses and suppression costs topping $100 million. In a strange example of natural cause and effect, the smoke and heat from large forest fires generate atmospheric conditions that produce even more lightning strikes, which cause even more wildfires.

For those suburban dwellers who think their sprawling communities represent more security than isolated houses in the country, there is bad news. According to some scientists, the increasing amount of urban structures clustered together often function as heat islands, and in some areas—Atlanta, Georgia, in particular—the added heat triggers a greater number of local thunderstorms. With these storms come more lightning strikes, and both the storms and the strikes generated from them are mainly increasing around city fringes, in the surrounding suburbs.

—Kim Long

RIGHT: *Originally reported by the local Knoxville, Tennessee media as a tornado, meteorologists later concluded that this storm was what storm chasers call a "scud bomb." While they can look like tornadoes or microbursts, scud bombs are actually clouds created by thunderstorm outflow lifting saturated air, which, in this case, was just above the ground. The severe thunderstorm that served as the source of outflow for this rare looking scud bomb was dumping more rain more than five miles away. Photo: Terry Mosher*

FOLLOWING PAGES: *A ghostly white tornado writhes its way out of the sky near Emmetsburg, Iowa, on June 11, 2004. This tornado was part of an outbreak of more than 40 tornadoes spawned by a potent late-spring weather system that traversed portions of northern Iowa and southeastern Minnesota. Photo: Gene Rhoden*

AMERICA'S WINDY CITIES

Chicago goes by many nicknames, among them "hog butcher to the world," "city of industry," and, in the poet Carl Sandburg's famous phrase, "city of big shoulders." Perhaps its best known nickname, however, is "the windy city," a sobriquet that owes its origins not to the weather—though, as anyone who knows Chicago can tell you, the winds blow fiercely off Lake Michigan at every time of the year—but instead to the city's longtime role as a venue for political conventions, where orators fill the air with windy words.

The average annual wind speed in Chicago is only about 10 miles per hour, but the wind is often a little brisker closer to the edges of Lake Michigan. March and April are traditionally the windiest months for this city. At different times of the year, storms may produce gusts of more than 50 miles per hour. In 1952, a November gale produced Chicago's record wind speed, 60 miles per hour.

America's true windy cities lie elsewhere, scattered across the country. Among the windiest are the great seaports of Boston and Providence, where cold winds from the North Atlantic collide with warmer winds spilling off the continental landmass to produce crying gales. One Boston suburb, Blue Hill, clocks an average year-round wind speed of 15.4 miles per hour. Similar figures come from Milwaukee and Des Moines making them far more appropriate to bear the moniker Windy City than nearby Chicago. And higher numbers still come from the towns and cities that lie along the Gulf of Alaska, such as Anchorage and Cold Bay, where the howl of the North Pacific wind accompanies the calls of blue whales and polar bears.

City residents experience higher wind gusts than their suburban and country counterparts because skyscrapers like these in Chicago off Lake Michigan make ideal wind tunnels and can turn the slightest breeze into a hurricane. Photo: Vic Bider/Getty Images

Other windy cities lie on the Great Plains, where, at all times of the year, cold winds blow straight down from the Arctic, unimpeded by mountains. The people of Great Falls, Montana, stand up straight not because of pride in their hometown—or so local legend has it—but because the constant winds keep them from slouching. Oklahoma City regularly clocks near-cyclonic winds, even in times when other pans of the Plains are relatively calm. Wichita, Kansas, and Cheyenne, Wyoming, are frequently battened by gale-force windstorms. And, ranchers will tell you, in West Texas the cows fall over in those rare moments when the wind is not blowing down from Alberta.

The windiest place in the lower 48 lies far from cities. Mount Washington, New Hampshire, is a forbidding granite spur high among the Appalachians, commanding the Presidential Range. Although not high by world standards, Mount Washington is a haven of howling winds and monstrously cold temperatures, and furious storms and hypothermia take their toll on human visitors in all seasons. The summertime temperature may approach 90 degrees Fahrenheit at Mount Washington's base, but the wind chill may take it down to freezing at the summit, for which reason hikers are warned to carry cold-weather gear even on the Fourth of July. The average wind speed on this peak is 35.4 miles per hour, and the amazing, all-time historical gust clocked in at 231 miles per hour.

Wherever you live, you're likelier to experience higher wind gusts in cities than in the suburbs or countryside, even though the average wind speed in cities is generally lower than that in open country. This is not some trick of nature to punish urbanites, but a simple fact of physics. A city street, lined with tall buildings and paved with smooth asphalt, makes an ideal wind tunnel, one that can amplify the slightest breeze into a hurricane. Not only that, but wind speed increases with height above the ground. On a city street with a wind speed of 30 miles per hour, it can be blowing at more than 60 miles per hour at the top of a skyscraper. In Chicago, the combined effects of breezes blowing in off Lake Michigan and the high, steep walls of skyscrapers can make the city's modest winds feel more like a nasty force of nature.

—Gregory McNamee

KILLER TWISTERS

In an average year, between 700 and 1,000 tornadoes are reported in the U.S., but most of these are small, weak, and cause little damage. Fewer than 20 percent of the tornadoes in an average year are violent and destructive; about 100 people are killed by twisters annually. In 1932, there were 151 tornadoes and 394 people died. The worst year for twisters was 1925, when 794 people died, 689 in a single tornado. The most destructive tornadoes, killer storms with winds in excess of 110 miles per hour (177 kilometers per hour), occur most often in the Midwest. Nebraska and Kansas are part of this "tornado belt," which extends from eastern Colorado east to Indiana, and from the Gulf states north to the Dakotas.

—Doris Grazulis

LEFT: *Debris flies from this violent F4 tornado after having just destroyed a farmhouse near Manchester, South Dakota. A scientific weather probe was deployed at the site of the farmhouse 70 seconds before the tornado struck. The probe measured an astounding pressure drop of 100 millibars in the center of the vortex. A survey of the damages supported estimated wind speeds in the range of 200 miles per hour. Photo: Karen Leszke & Gene Rhoden/Weatherpix Stock Image*

RIGHT: *The white house in this image was fortunate to have been spared by the F3 tornado looming overhead, which developed just south of Mulvane, Kansas, on June 12, 2004. As demonstrated by the flying debris, the house behind this one was not so fortunate. Estimated wind speeds between 180 and 200 miles per hour leveled the neighboring farmhouse and mangled the 1969 Mustang parked in its garage. Although it dissipated only one minute after forming, this was the strongest of six tornadoes to touch down on this day. Photo: Eric Nguyen*

TORNADO TYPES

Huschke's Glossary of Meteorology defines a tornado as "a violently rotating column of air (in contact with the ground) pendant from a cumulonimbus cloud . . ." To many people, it makes no difference how the tornado formed. It still causes damage, and threatens life and property. But to meteorologists, knowing how tornadoes form can help improve our early warning systems, and save lives. Until recently, scientists felt that most tornadoes were spawned from an intense thunderstorm called a supercell. But Doppler radar, vertical wind soundings, surface data, field observations, and other data have changed their thoughts. Now we know that tornadoes form in a variety of ways. Here are some of the more common types of tornado formation:

SUPERCELL TORNADO: These are typically the most intense tornadoes. Supercell thunderstorms develop when there is a favorable vertical "profile" of wind speeds and directions in the atmosphere. A clockwise change of wind direction and increase in wind speed with height creates invisible horizontal spinning "tubes" in the lower atmosphere. These spinning tubes become tilted into the vertical by thunderstorm updrafts. Through this tilting and stretching of the spinning tubes, the updraft of the thunderstorm begins to rotate about a vertical axis. This rotating updraft appears on Doppler radar (which measures winds) as a mesocyclone. The mesocyclone aids in creating a favorable mid- and low-level wind field in the thunderstorm from which a tornado develops.

LANDSPOUT TORNADO: This is a relatively new term to meteorologists. Landspouts form from non-supercell thunderstorms, and occasionally resemble waterspouts, which are common near the Florida Keys. They usually form during the early stages of a thunderstorm, but are not associated with a rotating updraft or mesocyclone. Instead, wind shift lines in the lowest layers of the atmosphere create vertical spinning "tubes." These tubes normally don't spin very fast until an intense convective updraft develops above them and stretches the tubes upward. When the spinning tube stretches, it spins faster, and the landspout forms. Typically, landspouts are weaker than supercell tornadoes, and are most frequent along the ocean coasts and east of the Front Range of Colorado. Landspouts in drier climates may appear as nearly transparent rotating columns of dust, with only a short funnel extending down from the cloud base.

On August 17, 1999, a massive supercell moved slowly eastward across western Nebraska and spawned several tornadic systems. These particular landspouts near Oshkosh were produced by a spectacular wall cloud that also generated very heavy rain and pea-sized hail. Photo: Gordon McNorton

WATERSPOUT: These are similar to landspouts, but form over the water. Although common over the warm waters of the Gulf Stream and the Florida Keys, they sometimes form when cold air masses pass over the relatively warm Great Lakes in late summer and autumn. Waterspouts can move onshore and cause tornado-like damage. Supercell tornadoes which pass over the water are also called waterspouts. Scientists clearly recognize the differences between them and true waterspouts.

GUSTNADO: These rotating columns of air are normally found along the leading edge of the cool outflow (gust front) preceding a thunderstorm. Typically, they are weak and last less than a minute. Some researchers argue that these may not be true "tornadoes" at all. They may be more akin to a swirl of leaves created by wind eddies near edges of buildings. Although gustnadoes usually do not cause severe damage, a few have been observed to cause tornado-like damage over a short and narrow path.

HYBRIDS AND OTHER TYPES OF TORNADOES: Meteorologists are still unsure of the causes of certain kinds of tornadoes. Some appear to be a combination of two or more of the types mentioned above. But the causes of other types, with names such as cold-air funnels and flanking-line tornadoes, still elude researchers. Although dust devils are rotating windstorms which on occasion cause damage, they are not associated with thunderstorms, and hence not classified as tornadoes. As long as questions remain unsolved, scientists will continue to probe the mysteries of these incredible wind monsters.

—*Gregory J. Stumpf*

LEFT: *Just before sunset a fast-moving storm darkens the skies on a June evening near Poplar, Montana. Short-lived but powerful, the storm produced brief but strong wind gusts, heavy rain, light hail, lightning, and after its passing, rainbows. This image depicts the leading edge of a cold front passing over the high plains. The clouds are in a transitional state as the cold front converges with the 100 degrees Fahrenheit air of the western plains. The varied and fast-changing winds and eddies show the flow patters and fluid motions of the storm. Photo: Bob Firth*

RIGHT: *Stormy skies unleashed two tornadoes on the beaches of South Carolina on July 6, 2001, causing an estimated $8 million in damages. Thousands of people watched from their hotel balconies, homes, and cars as the tornadoes cruised down the "Grand Strand" during the busiest tourist week of the year. No one was killed. Photo: Paul R. Donovan*

MONSTER DUST STORMS

When most Americans think of dust storms, images of the Dust Bowl in the 1930s come to mind. For most of that decade, a drought affected significant parts of the country, triggering large amounts of soil erosion, most of it into the air. At certain times, such as April 14, 1935, known as "Black Sunday," the clouds of dust and blowing topsoil were so dense and massive that the atmosphere over entire states was saturated, blocking out the sun.

These days, large dust storms are becoming news again, although this time they are mostly originating in Africa, not North America. A huge belt of traditionally arid terrain stretches across the northern half of this continent. Afflicted by a long-running drought and increasing human activities that disturb the top layer of soil, this region is pumping out more and more giant dust storms. In the city of Nouakchott, for example, on the west coast of the country of Mauritania, situated in northwestern Africa, about 5 to 10 dust storms were typical for an average year in the 1960s. By the 1980s, the frequency increased to about 80 per year.

The massive storms that plague western Africa are driven west by prevailing winds. Seasonal conditions are at their driest in the first half of the year; it is during this period when most of the largest storms occur. Because of seasonal wind patterns, the storms are most likely to follow a track to South America or Central America early in this dry season. Later in the season—June and July—the storm tracks often shift farther north, where they intersect the Gulf of Mexico, the Bahamas, the Florida Keys, and the southeastern coast of the United States.

Dust kicked up into the air during these storms adds up to a lot of dirt in the course of a year. Estimates put the total volume in the hundreds of millions of tons, and a single storm can carry million of tons. Unlike volcanic dust, however, most of this material does not end up high in the atmosphere. Most windblown dust stays in the troposphere, the lowest part of the atmosphere extending from sea level to about 10 miles, and doesn't stay suspended for long, a week at the most. Most of this blowing material drops close to its source, creating drifts, piles, and general havoc. But the material that remains airborne can end up going for quite a long ride, ending up affecting the ecologies on both land and in the ocean.

*Newly arriving Sudanese refugees can only watch as a huge sandstorm approaches the Ouri Cassoni refugee camp just outside Baha'i, Chad. This was most likely a convective sandstorm typically associated with thunderstorm outflows and microbursts. Convective sandstorms are less predictable than nonconvective storms and are usually identifiable only a few minutes before they hit. If caused by a microburst, one may last for only a few seconds, whereas one caused by a macroburst can last a few minutes. One caused by a weak depression can last hours. Convective sandstorms mostly occur in late afternoon in the spring and summer months. Photo: Jahi Chikwendiu/*The Washington Post

The particles of soil that are part of dust clouds are often rich in nutrients. There is evidence that for thousands, if not millions, of years, this has been a significant benefit. In the Bahamas, for example, red soils favored for cultivation are accumulations of dust from African storms over long periods of time. And this is not the only such beneficial relationship on the planet; a similar nutrient-rich deposit has been going on for a long time in the rainforests of the Hawaiian Islands thanks to dust storms in Asia. Here, the dust has been carried more than 6,200 miles, originating in the Taklamakan Desert in China.

Unfortunately, this beneficial relationship is only part of the story. With the findings of research that has been going on since the 1960s, it is becoming evident that there is also a more deadly symbiosis. The dust blown out of Africa contains fungus spores, iron, phosphorous, sulfates, and other compounds. One fungus in particular, a species of *Aspergillus*, can cause lung infections in humans and the demise of marine animals such as the grazing sea urchin, sea fans, and certain kinds of corals. In fact, there is a direct correlation between major dust-storm events in Africa and peak die-offs of corals off the coast of Florida and elsewhere in the Caribbean. The inorganic compounds found within African dust have a different effect. They act as fertilizers and may contribute to the occasional build-up in populations of sea-based microscopic organisms; when population explosions occur for these organisms, red tides are a result. The dust may also be one factor associated with oxygen depletion in estuaries and localized areas of the ocean, killing off most kinds of aquatic life. In addition, one or more of these dust-born contaminants could be linked to regional population declines in amphibians. Although it is mostly lightweight soil particles that are blown across the ocean, the power of the dust storms can also carry along larger items, such as the African desert locusts (a kind of grasshopper), which were dropped on Trinidad after a dust storm in 1989.

As scientists explore more complex climate conditions, the big dust storms have become a new focus of interest. The particles carried along by the wind not only have direct consequences for human and animal lives along their route, merely being suspended in the air has another effect as well. In regions near to where the dust originates, this may include a decrease in rainfall, the result of the dust stabilizing local movements of air, keeping it from rising and moisture from condensing. And even though water drops normally form around dust particles in the air, when the concentration is high the effect is the reverse and fewer drops form. Some sizes of dust particles in the air can also reduce the amount of solar energy reaching the earth's surface, scattering the radiation and reducing its warming effect. Just this kind of action from volcanic dust spewed into the air by the eruption of Mount Pinatubo, 55 miles northwest of Manila, Philippines, in 1991 is credited with dropping the mean global temperature by about one degree Fahrenheit. But the effect of some kinds of dust particles in the air can also be the reverse, contributing to surface warming by absorbing infrared radiation that would otherwise travel up and out of the atmosphere.

Although the African desert may be a major culprit in producing large amounts of dust in the atmosphere, other continents also play a role. Construction and agricultural expansion in Asia, for example, have made conditions ripe for an increase in soil erosion, and the results may extend far to the east from that continent. In North America, an increase in soil erosion in the Southwest is also noted, pumping more dust into wind patterns that carry it as far as the Gulf of Mexico and Florida.

—*Kim Long*

This monster dust storm swept across fields four miles north of Colby, Kansas, and lasted for around two hours, with wind gusts up to 86 miles per hour and zero visibility. Two-inch-tall corn growing in nearby fields turned black from static burn due to the dirt and electricity in the air. Kansas state senator Stan Clark of Oakley, Kansas, lost his life on Interstate 70 as a result of the storm. Photo: David Pabst

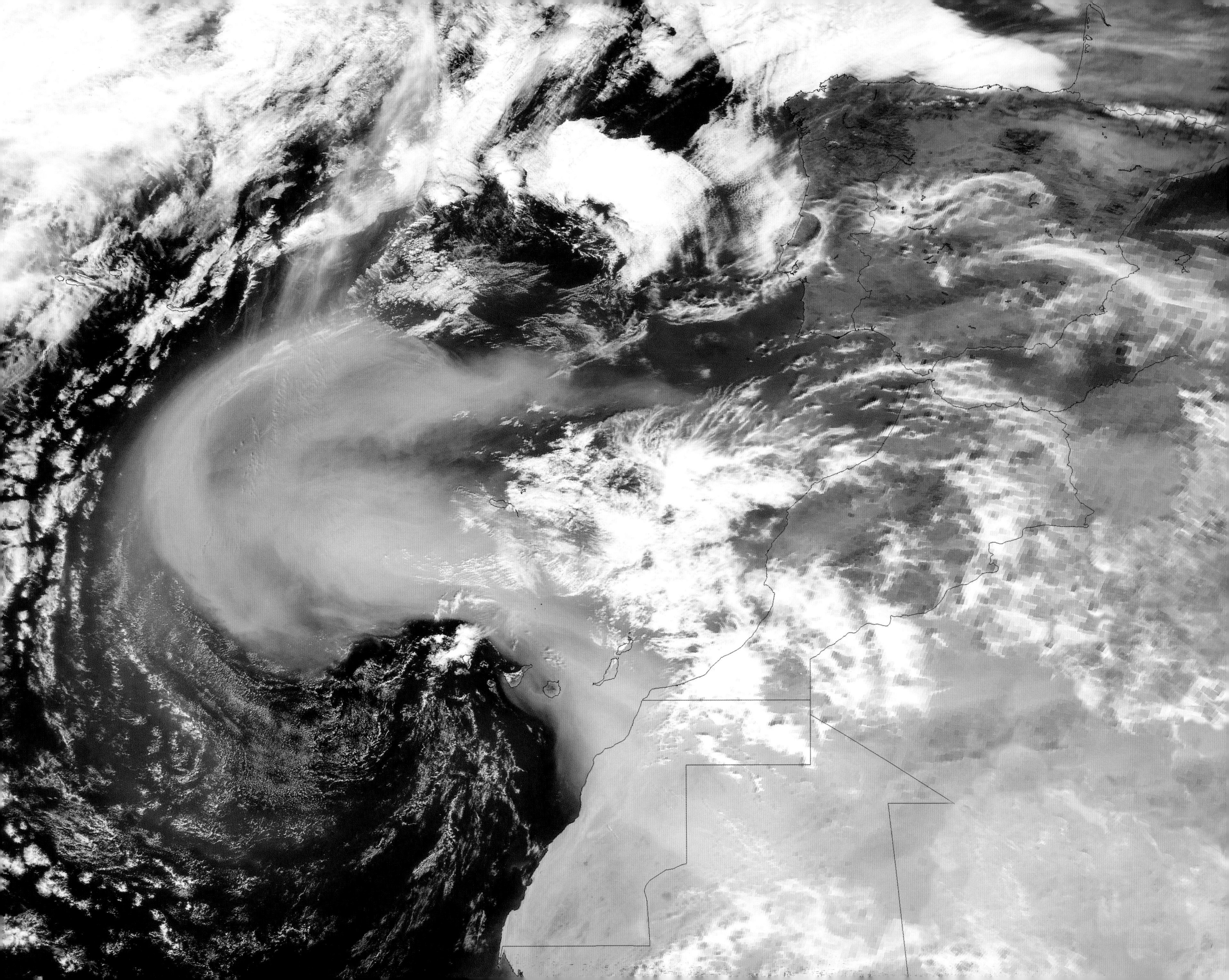

LEFT: *A massive sandstorm blowing off the coast of the northwest African desert blankets hundreds of thousands of square miles of the eastern Atlantic Ocean with a dense cloud of Saharan sand. These storms and the rising warm air can lift dust 15,000 feet or so above the African deserts and then out across the Atlantic, many times reaching as far as the Caribbean. Recent studies have linked the decline of the coral reefs there to increasing frequency and intensity of these storms. Other studies show that Sahelian dust may play a role in determining the frequency and intensity of hurricanes in the eastern Atlantic Ocean. Photo: Orbimage*

RIGHT: *Ground temperature was nearing 100 degrees Fahrenheit when this large dust devil appeared in the Kalahari Gemsbok of South Africa. Dust devils occur most frequently on clear, hot days in the southwestern United States and in other desert areas around the world. Intense heating at the earth's surface produces sharp temperature contrast between the ground and the very lowest levels of the atmosphere, causing rapidly rising warm air to produce a mini-low pressure system. As the hot air rises, dust is picked up from the ground by a twisting motion caused by varying winds blowing into the system, or from frictional forces resulting from changing terrain. Photo: Stan Osolinski/Dembinsky Photo Associates*

WATER TWISTERS

Waterspouts in the Gulf of Mexico like this one near St. Petersburg, Florida, draw up large columns of water as well as the associated inhabitants trapped in the updraft. Occasionally, waterspouts will move inland where they become tornadoes, releasing their load of fish and other sea creatures. The peak season for waterspout viewing in the Florida Keys is during the late summer months. Although usually smaller and weaker than land tornadoes, waterspouts have shown speeds of up to 190 miles per hour and should not be taken lightly. The best escape from a waterspout is at a right angle from its path. Photo: E. R. Degginger/Dembinsky Photo Associates

In some parts of the world, waterspouts are a common sight. These twisting funnels of water are technically known as "convective vortices" and are similar to dust devils, which form over land. Most of the time, waterspouts occur on bodies of water with high surface temperatures, usually in tropical regions. The big spouts may generate vortices of 100 miles per hour or higher. In the ocean around the Florida Keys, as many as 500 waterspouts form every summer, some of them of the larger variety. Farther north, drivers crossing the Tampa Bay Bridge are frequently entertained by the sight of these dancing funnels during the summer months. Waterspouts are also sometimes observed in the Great Lakes, along the Gulf Coast, and in the southern part of the Pacific Coast. Several have also been noted on the Great Salt Lake in Utah.

Waterspouts form under a variety of conditions, but are mostly associated with the growth phase of cumulus congestus clouds. As a line of these clouds forms, waterspouts often follow, sometimes appearing in groups. A waterspout, visible because of the spinning spray of water that it is composed of, usually extends from the surface of a body of water up to the base of the cloud above it, typically only a few thousand feet above the surface. Often, the funnels formed by the smallest, most common waterspouts twist and snake unevenly across this distance. Depending on the amount of water in the funnel, all or part of the funnel may be observable. In general, the closer to the water, the more visible the funnel. As they form, waterspouts often kick up a circular movement of water near the surface, making a ring of spray at the base of the spout, which appears larger than the funnel that extends out of it. This feature is called the spray sheath. Because of long lines of visibility on the water, observers are often able to see the entire life cycle of a waterspout, beginning with the first signs of rotation on the surface. Waterspout funnels often grow in size as they "mature," and may form and reform more than once before dissipating. As spouts decay, they become thinner and fainter in appearance.

Most waterspouts are small, with thin, ropelike funnels that may only be a few feet in diameter, but sometimes much larger spouts are generated. The big ones feature higher rotation speeds and larger funnel diameters, sometimes as large as tornadoes. And just as tornadoes pack enough power to pose a threat to people and property, so too can large waterspouts produce damage and death, although this is rare.

LAKE TAHOE WATER TWISTERS

On Lake Tahoe, high in the mountains in northern California, waterspouts are not common. In fact, waterspouts are so rare in mountain lakes that only a few have ever been recorded anywhere in the world. On September 26, 1998, however, the weather at Lake Tahoe (elevation 6,200 feet) featured a unique set of conditions ripe for the production of waterspouts.

These high-altitude twisters occurred between 7:00 and 9:30 A.M., forming and moving across the northern part of the lake. Several conditions combined to produce this event, although the exact set of circumstances is not known. One contributing factor may have been the surface temperature of the water, in the low 60s, but close to its peak temperature for the year in this high, cold environment. In these surroundings, the surface temperature rises due to the absorption of heat from the sun. At the same time, cyclonic circulation in the atmosphere of the region helped converge moisture at a relatively low altitude, related to an upper-level low, forming a line of cumulus congestus clouds and ideal conditions for the creation of waterspouts. Another potential factor may have been surface winds channeled by the local terrain around the shore of the lake.

A waterspout at Lake Tahoe is not only rare, the fact that there were at least six of these twisters observed makes it even more unusual. Stranger still, one of the spouts was of unusual size. This large funnel of vapor gradually expanded to be more than 250 feet in diameter and could have been a serious threat if it had encountered any boats or structures on shore. Before this event, only about 12 waterspouts had been reported on Lake Tahoe in the previous 30 years.

—*Kim Long*

Waterspouts over mountain lakes such as this one over Lake Tahoe are extremely rare. This waterspout occurred on the morning of September 26, 1998, approximately two miles off the Cal-Neva point. The sun line on the water is very close to the imaginary California-Nevada border that extends across Lake Tahoe. Tahoe Vista, California, is in the foreground and the Sierra Nevada mountain range is in the background. The depth of the lake in the area of the twister is well over 1,000 feet—one of the deepest parts of the lake. The diameter of the twister was about 275 feet with a spray of 500 feet across its base. Photo: Gary Kaufman

LEFT: *A rare waterspout in the Virgin Islands was formed when a cold front from the United States moved in and hit the warm tropical air. These waterspouts were coming up the East Gregory channel between Hassle Island on the left and Water Island on the right. St. Thomas is in the foreground. Photo: Gary Felton*

RIGHT: *Waterspouts are a dramatic sight during summer months in the Florida Keys as they form over water along the outer edges of heavy cloud build-ups. These large twin waterspouts formed in early summer just offshore Summerland Key, Florida. Lasting more than 30 minutes, the huge funnels put on a spectacular show for man and cormorant (in foreground) as they repeatedly rose and descended and crossed in front of each other. Photo: Ken Araujo*

THE NATURE OF HURRICANES

LEFT: *Hurricane Frances hits near Juno Beach, Florida, with hurricane force winds on September 4, 2004. Photo: Warren Faidley/Weatherstock*

ABOVE: *On August 28, 2005, while in the Gulf of Mexico, Hurricane Katrina reached Category 5 status, with maximum sustained winds near 175 miles per hour as it approached land the next day as a Category 3 storm. Photo: NOAA*

Hurricane. The word, derived from a Carib Indian phrase, means nothing more than "big wind." A hurricane is a big wind indeed, technically defined as a strong tropical cyclone with sustained winds of more than 74 miles per hour. But how that tropical cyclone got to spinning in the first place: well, that's a story, and one that scientists are still working on understanding.

In order to make a hurricane, nature has to bring together several disparate elements. First, there has to be an ocean, and at the surface of that ocean, the water must have attained a temperature of 82 degrees Fahrenheit (28° Celsius). Then there must be a clash of hot air and cold air, just the sort of thing that occurs when the cool waters of West Africa's Gulf of Guinea flow northward past the coast of Mauritania and Morocco and there meet the searing winds spilling off the Sahara Desert.

Air meets water, and water meets land, forming choppy waves that bounce back off the shoulder of the African continent. Those waves fuel the initial spin that sends a column of hot air, now moisture-laden, whipping upward. As this air rises, it cools rapidly, causing clouds of condensation to form and rain to fall. Given sufficient energy—and hurricanes are great engines that gather energy from near and far—the column of rising air will begin to turn cyclonically, moving with the prevailing winds and pulling up more and more warm water.

The power of the column builds. The barometric pressure drops, whence the term "tropical depression." The depression begins to generate ever more intense winds, while lightning crackles from the clouds and tall, long waves form below the growing storm system. The center of the vortex, the fair-weather eye of the storm, is calm, but all around it is weather that is anything but fair. As the storm nears land, the winds push water toward the shore. This "storm surge," as it is called, joins with the normal tides, creating the hurricane storm tide, a swell of water that is 15 or so feet higher than the normal water level and that causes trouble all on its own, quite apart from the furious winds that drive it.

Energy systems, the laws of thermodynamics advise, have a tendency to fall apart in time. Sometimes a hurricane can collapse under its own weight, hit cold water in the middle of the Atlantic, and go nowhere.

One of the few known pictures of Hurricane Andrew as it hit Dinner Key, Florida, on August 24, 1992, with winds of over 160 miles per hour. Photo: Warren Faidley/Weatherstock

Sometimes it will make landfall, bash around for a while, and then sputter out. Sometimes, though, a hurricane will make land and stay alive, skipping across humid islands and peninsulas such as Cuba and Florida to find more warm water beyond.

Thus it was with the most devastating and expensive hurricane to hit America in its recorded history—though far from the most powerful. Hurricane Katrina began life in the twinkling of a storm's eye off the coast of northwestern Africa, gained momentum as it slowly lurched westward, and rose to howling fury when the warm waters off the Bahamas refueled a system that, to that point, seemed likely to fall apart at any moment. Instead, it banged into Florida, wandered across the low, flat, and hot southern extremes of the peninsula, and reached the still warmer waters of the Gulf of Mexico.

What happened next is seared into the nation's memory: more than 1,800 lives extinguished, billions of dollars lost, whole cities destroyed or badly damaged, and the unsettling knowledge gained that even with ample warning, a sophisticated postindustrial society can be easily shattered by the force of a storm.

Hurricanes have their place in the world, like all of nature's creations. Although they can be disruptive and destructive, they can even bring some good, at least from the big-picture point of view. They can restore bleached coral reefs, dredging up cold water from the depths of the ocean and bringing new nutrients to algae-deprived reef formations. And, scientists have found, tropical storms and hurricanes cause surface-water cooling when they pass, having removed so much warm surface water in the process. The cooling effect can last for up to 40 days—a great boon for overstressed oceanic ecosystems in these days of global warming.

But when those 40 days are over, another hurricane may lie in wait. Thanks to computers and other technological advances, we have ever more data and knowledge about the nature of hurricanes at our disposal. But how are we to use that information to save lives and property? We'll need to work on that, and by all indications, we'll have plenty of opportunities to do so.

—Gregory McNamee

FLORIDA: A HURRICANE MAGNET?

Floridians are used to storms, to wind and rain and thundering seas. Those storms are legendary, the sort of thing that Floridians remember, recount, and compare notes on. They're part of what makes the Sunshine State the—well, historically the hurricane capital of the country, which the state's byname fails to mention.

Four major hurricanes slammed into the state within just six weeks in 2004. In 2005, Hurricane Katrina arrived at Miami as a Category 1 storm, skipped across the lower part of the state, and gathered its monstrous force over the Gulf of Mexico. Two months later, Hurricane Wilma, a Category 3 storm hit southwestern Florida, causing tremendous wind damage and flooding, and costing $16 billion in damages and an estimated 35 deaths.

It's not that the lord of the hurricanes has it in for the low-lying peninsula. Rather, the Sunshine State happens to sit dead in the midst of a global locale exceedingly fertile for the great storms, which start life on the other side of the Atlantic Ocean and make a beeline straight for Florida.

A hurricane needs warm, moist air to develop and grow, and it finds it in the waters of the West African tropics. As it gathers force, it begins to spin cyclonically, crossing the Atlantic Ocean in a slow counterclockwise spin that finds Florida thanks to the inviting pull of the Gulf Stream, a 40- to 50-mile-wide "river" of warm water that begins in the Gulf of Mexico, loops around the southern tip of the Florida Peninsula, and then runs parallel with the eastern United States coastline before shooting out into the northern Atlantic Ocean.

Offering a fresh supply of warm water to keep an ocean-crossing hurricane alive, the Gulf Stream is a magnet for such great storms. The Gulf Stream feeds that hurricane very well indeed, refuels it, and gives it new strength. Once the storm moves over Florida, it immediately falls ill and begins to die. However, if it crosses the peninsula and makes it to warmer, moist Gulf Stream air in the Gulf of Mexico, then it may lurch back into life—as was the case, notoriously, with Hurricane Katrina.

Are the recent hurricane seasons typical or anomalous for Florida? To make a truly accurate statement, meteorologists and statisticians will need more years of data, but the record now suggests that the Gulf Stream makes it an obvious target for those great African-born storms. Yet, fierce and unforgiving, the storms that are still fresh in memory have put many Floridians to wondering just how safe their paradise really is.

—Ed Darack and Gregory McNamee

MOST EXPENSIVE U.S. NATURAL DISASTERS

According to the National Oceanic and Atmospheric Administration (NOAA), which calls itself the "nation's scorekeeper" of weather-related statistics, more than 70 weather events reached the billion-dollar threshold between 1980 and mid-2007—and all but seven of those occurred after 1988, which suggests a pattern of growing storm intensity and growing costliness alike. The total damage of those storms, adjusted for inflation in 2007 dollars, amounts to about $640 billion—a sum equal to the projected federal defense budget for 2008.

Following are the most costly storms in recent years, with reported total damages and costs. The adjusted values for 2005, the year of Hurricane Katrina, are in parentheses, and the rankings are weighted by those figures.

1. **August 2005:** Hurricane Katrina
   Gulf of Mexico coast, especially Louisiana, Missisipi, and Alabama—$125 billion
2. **Summer 1988:** Drought and heat wave
   Central and eastern states—$40 billion ($66 billion)
3. **June–September 1980:** Drought and heat wave
   Central and eastern states—$20 billion ($47.4 billion)
4. **August 1992:** Hurricane Andrew—$22.3 billion ($31 billion)
5. **Summer 1993:** Flooding in the Midwest—$21 billion ($28.4 billion)
6. **September 2005:** Hurricane Rita
   Texas coast—$16 billion; Hurricane Wilma, Atlantic coast of southern Florida—$16 billion
7. **August 2004:** Hurricane Charley
   $15 billion ($15.5 billion)
8. **September 2004:** Hurricane Ivan
   Gulf of Mexico coast, especially Florida—$14 billion ($14.5 billion)
9. **September 1989:** Hurricane Hugo
   $9 billion ($14.2 billion)
10. **Summer 2002:** Widespread drought and heat wave
    Affecting 30 states—$10 billion ($10.9 billion)

It is estimated that within the last two centuries hurricanes have been responsible for the deaths of about 1.9 million people worldwide.

HURRICANES & UNITED STATES HISTORY

LEFT: *Flooded roadways can be seen as the Coast Guard conducts initial Hurricane Katrina damage assessment in New Orleans. Photo: Kyle Niemi, courtesy of U.S. Coast Guard*

ABOVE: *Mobile, Alabama's local newspaper warns its readers of the imminent onslaught of Katrina, the Category 5 hurricane which struck Mobile on August 29, 2005, with a 15-foot storm surge and winds over 100 miles per hour. Photo: Warren Faidley/Weatherstock*

Study a map of the United States. Consider the sheer size of the North American continent, the thousands of miles of coastline along the Atlantic Ocean and the Gulf of Mexico, and the country's location relative to the tropical waters of West Africa—all accidents of geography that put the nation squarely in the path of hurricane after hurricane, year after year.

Long before hurricanes had names, long before there was any kind of weather forecasting or recording service, hurricanes were changing the course of American history. One severe hurricane actually aided the American colonies in their struggle for independence. Known in history as the "Great Hurricane," it tore through the Windward Islands and Puerto Rico in October 1780, leaving an incredible 22,000 fatalities.

The storm dealt a heavy blow to the Royal Navy, diverting English warships away from American waters. Especially hard hit was a convoy bringing reinforcements to America, with the loss of many ships and veteran soldiers. That devastation played a considerable role in the British decision not to challenge the French blockade of Yorktown in October 1781, which forced the British surrender there and paved the way for American independence.

Earlier, hurricanes wreaked havoc with Spanish enterprises in the New World. In the 16th and 17th centuries, British privateers forced Spanish treasure-laden ships to assemble in convoys for their return to Spain. Often setting sail during the hurricane season, many ships were lost. Some convoys were said to have carried a billion dollars in treasure.

It is estimated that there was a hurricane-induced shipwreck for each mile of the Florida coastline. These losses were undoubtedly a major factor in thwarting Spain's drive to dominate the Atlantic coast of North America. The French suffered, too; in 1565, a French fleet was destroyed by a hurricane off Fort Caroline, between St. Augustine and Jacksonville. As a consequence, peninsular Florida remained Spanish territory for the next two centuries.

In 1815, the storm called the "Great September Gale" struck New England, causing major damage to the great fishing fleets that sheltered along the Massachusetts coast. Scores of New Englanders died, an event that figures in the memories of the old-time sailors in Herman Melville's novel *Moby-Dick*. But that storm, and indeed most of the hurricanes of the nineteenth century, was modest compared to the one that destroyed Galveston, Texas, in 1900.

For the next century, the Galveston storm would rank as the nation's worst natural disaster. Other terrible storms gave Galveston a run for its money, though. In 1926, a powerful hurricane brought the Great Depression to Miami several years before it began across the rest of the nation. In 1928, a hurricane broke the levees around Lake Okeechobee, drowning hundreds south of the lake. In 1935, the Flagler Railroad in the Florida Keys was destroyed, never to be replaced, and hundreds were killed in the strongest hurricane ever to strike the United States to that time. More than 600 died in the New England hurricane of 1938, the most powerful to strike there since 1815.

In 1969, Hurricane Camille, only the second Category 5 hurricane to make landfall in the United States to that date, laid waste to the Mississippi coast and traveled far inland. Hurricane Hugo struck hard in 1989, and then came the storms of the 1990s, foremost among them Andrew—but also including Bertha, Gordon, Irene, Mitch, Floyd, and other innocuously named but deadly tempests. They would pale in comparison to the ones that would arrive in the early years of the twenty-first century: Ivan, Dennis, Jeanne, Rita, Emily, and Katrina.

The last, which made landfall in August 2005, destroyed huge portions of the central Gulf Coast, devastated the cities of New Orleans, Biloxi, and Mobile, and cost more than $135 billion in damage—and the loss of more than 1,800 lives. Two years afterward, those costs were still being analyzed and revised ever upward. What is certain is that Katrina was the most expensive, most damaging storm in American history.

Hurricanes are constants of our past, present, and future. As more and more Americans come to live within the regions of the country most susceptible to damage by them, it becomes ever more urgent that we make our best efforts to prepare ourselves for these powerful storms.

—John Hope and Gregory McNamee

Huge waves hit the coast of North Carolina as Hurricane Isabel approaches Atlantic Beach on September 19, 2003. Photo: Warren Faidley/Weatherstock

WATER

RAIN, ICE, SNOW, FOG, HAIL, AND CLOUDS

It is indeed a strange and versatile molecule. Two atoms of hydrogen and one of oxygen, it is the rare substance that exists as a gas, solid, and liquid at terrestrial temperatures and pressures.

Our planet, it has been remarked, should be called Aqua, not Earth, given that more than 75 percent of its surface is water in either liquid form (the hydrosphere) or solid (the frozen cryosphere.) We have 326 million cubic miles of H_2O—seems like plenty until you realize that barely 3 percent of it is fresh and that over a billion humans daily struggle to find potable water.

Our atmosphere contains far less than 1 percent of the total water, but does amazing things with its meager share. The winds transport water mass within the hydrological cycle, the continuous pageant of evaporation to clouds, then precipitation, runoff into streams, lakes, and oceans and back to evaporation. Replenishing surface water by precipitation (ignoring the scant contributions of frost, dew, and fog droplet capture by plants) exhibits considerable versatility: mist and drizzle, rain, both regular and freezing on contact, ice pellets, graupel and hail, and the dazzling diversity of six-sided crystalline forms in every snow storm.

LEFT: *Few settings can be more spectacular than Arizona's Monument Valley, especially when coupled with a magnificent thunderstorm. This early October image was taken over "The Mittens" in Monument Valley Navajo Tribal Park. A rain shaft and rainbow to the left, with the sun's crepuscular rays (caused by sunlight and cloud shadows passing through airborne dust particles) to the right, combine to make this extraordinary scene. Photo: Curtis Martin*

FOLLOWING PAGES: *A monsoon storm over Benson, Arizona, is captured in August 2001. The Sonoran Desert encompasses 120,000 square miles in the Southwest United States. The monsoon rains typically arrive in July after the desert has been scorched by temperatures that routinely reach 110 degrees Fahrenheit. Weather extremes in the form of flash flooding, frequent lightning, dust storms due to the thunderstorm outflow, high winds, and microbursts are common during the monsoon season between July and September. Photo: Susan Strom*

A single raindrop rounds up a million cloud droplets, and a million fellow droplets congeal into a golf-ball-sized hailstone. Precipitation is an element of extremes. A Maryland gauge once recorded 1.25 inches of rain in 60 seconds, while gauges in Chile's Atacama Desert have waited 60 years to see so much as a drop. The failure of expected rains has decimated great civilizations, while in modern times the 1988 drought was America's costliest natural disaster until Katrina. Yet only five years later much of that same region was inundated by massive flooding in the Missouri and Mississippi river basins. Floods remain the principal severe weather killer of Americans year in and year out.

Water vapor condensing in our skies creates sculpted cloud artworks as in the lens-like altocumulus lenticularis riding a mountain wave. The average citizen is compelled to look skyward when voluptuous mammatus clouds undulate beneath a thunderhead anvil. Seemingly granitic cloud mountains are rendered by thunderstorm updrafts. And delighted glider pilots soar on tubular waves in Australia's Morning Glory cloud. Subtler are the gentle filigreed ice crystal trails of cirrus clouds. But a cloud resembling Swiss cheese startles after penetrating aircraft shock supercooled droplets into precipitating snowflakes. That same conversion of supercooled cloud droplets into snow forms the basis of artificial rain and snowmaking. This decades-old technology is returning to prominence as water shortages plague a planet with depleting resources. Meanwhile, the heat and pollution of our cities and burning forests may inadvertently be affecting regional clouds, rainfall, and lightning in ways we are just beginning to decipher.

The planet's frozen water assets are melting at an increasingly alarming rate. While Arctic sea ice reverting to liquid state will not raise sea level (check ice cubes melting in a glass of ice tea), the runoff from mountain glaciers, Greenland, and the great Antarctic ice sheets may alter coastlines in ways barely thinkable, even with the visual aids of inconvenient truths. Even just warming the ocean causes water to expand and sea level to rise. Our human bodies are mostly water, and without water, what we know as weather would not exist. Water is a strange and versatile molecule—and the essence of life.

—Walter Lyons, CCM

MOUNTAINS: WATER TOWERS OF STONE

LEFT: *At 20,320 feet, Alaska's Mt. McKinley, also called Denali ("high one"), is the highest mountain in North America. The mountain's imposing structure is ideal for the formation of lenticular clouds. As moist air rises over McKinley, water vapor in the air is squeezed together to form lens-shaped clouds. When the heavier lens-shaped clouds sink, their moisture evaporates, leaving the cloud structure intact to be shaped by the prevailing winds. This process repeats itself to provide layers of moist, condensed air, separated by layers of dryer air, which appears as gaps between the clouds. Timberline on Mt. McKinley reaches only 3,000 feet. Its peaks are permanently snow-covered and are visible from more than 250 miles. Photo: Howie Garber/Wanderlust Images*

ABOVE: *El Niño moisture brings life to California's Anza-Borrego desert. Photo: Carol Simowitz*

It begins with a dry crack, a musket shot from some heavenly picket line, and then expands into a cannonade—boom, boom, and boom! Loud enough to rattle teeth. The sky above the Gulf of Policastro, on the coast of southern Italy about a hundred miles below Naples, is clear, but the Serra del Tuono, "the mountain range of thunder," is living up to its name on this early autumn day. The noise is a portent, and sure enough, from the east and the country's narrow interior come clouds that bubble up into a fierce, but brief, rainstorm. Soon, sheets of water come pouring down from the mountains and crashing into the wind-whipped Mediterranean Sea.

Geology and climatology conspire to make mountains important sources of water, helping to create and then release their bounty to the surrounding lowlands in regular and usually predictable patterns. Mountains are also responsible for much of the planet's water storage, tucking it away in underground springs and packs of snow and ice until it is needed. And it is always needed.

Mountains such as the Serra del Tuono create their own weather; they call up clouds and thunder from seemingly nowhere, as if calling for a drink of water. This is as true of mountains tucked away in the remotest corners of the desert as it is for those in the tropics, for mountains form natural impediments that milk the moisture from sinking, water-laden air masses while routing hot, dry winds upward in thermal columns. Anabatic and katabatic winds (upward-flowing and downward-flowing, respectively) surround most mountains, lending the airspace above the rocks a choppy quality that airline passengers know well. They also play a role in several aspects of continental weather patterns, including the so-called rain-shadow effect of the deserts of western North America and the autumn foehn of the southern Alps, a hot, dry wind that sometimes melts the first snows of autumn and imperils late-season mountaineers.

While mountains release the water they collect, they also draw on the stores of water that are naturally held within the rock. Just as mountains are laced with veins of ore, so too has volcanic action forced veins of underground water up to the highest summits, forming the springs and tarns (mountain pools or lakes) that give birth to so many of the world's rivers.

Located on the Navajo Reservation north of Winona, Arizona, this image shows the Little Colorado River in flood pouring over what has been named the Grand Falls of the Little Colorado. Heavy monsoon rains sent floodwaters flying over these 200-foot-high, 400-foot-wide cliffs at a rate of 27,000 cubic feet per second—impressive for a little riverbed that is normally bone dry. By comparison, Niagara Falls is approximately 167 feet high and flows continuously at an average rate of 20,000 cubic feet per second. Photo: Tom Brownold

Even in the most forbidding deserts, mountain springs do their work. In the Tassili Mountains, for instance, in the heart of the Sahara, stand ancient groves of well-watered cypress trees and small lakes full of fish, all thanks to the range's talent for storing water. In the mountains of southeastern California, wind caves deep within the rock shelter groves of palm trees long extinct elsewhere, providing water for dozens of species of plants and animals—including wayward humans. Water even seeps from the great monolith called Uluru, deep within the continent-wide Australian Desert—water that lends its name to the town of Alice Springs.

Another desert town that benefits from the presence of the great stone water tanks that mountains afford is Chinle, Arizona, at the mouth of Canyon de Chelly. The little town takes its name from the Navajo words meaning "where the water comes out," and if you follow that water, the long stream called Chinle Creek, through the deep, white-walled canyon complex and over the rugged escarpment called the Defiance Uplift, you will eventually find one of its sources in the Tunitcha Mountains, whose name means—in the local native language—something like "where much water comes from." The name fits perfectly. Not only do the often snow-clad mountains shed water into abundant creeks feeding into the distant Colorado River (which provides two-thirds of the surface water found within the entire Navajo Nation), but the top of 9,512-foot-tall Matthews Peak, the Tunitchas' highest point, is also dotted with natural springs and water holes. It is easy to see why the Navajo novelist Irvin Morris described the area as "an archipelago of well-watered islands."

In more temperate climates, the mountains are correspondingly generous, which is why many mountain chains there are pockmarked by man-made dams such as the Hetch Hetchy of California and the Three Gorges of China. The stores of water within the ancient rock in such places are augmented by rainfall carried in by seasonal winds, often in quantities many times greater than the rainfall of the neighboring lowlands. If a mountain is well forested or contains adequate natural storage facilities, this rainfall will percolate down to the lowlands at a more or less even pace. Where mountain areas have been heavily logged or lack reservoirs and water-draining meadows, there is always the danger that highland storms will lead to lowland flooding. This was the case during the Big Thompson flood of 1976 in Colorado, which saw extraordinarily heavy rainfall—more than 14 inches in six hours—rocketing down a steep, narrow canyon without side drainage, and killing 139 people along the way.

Water stored within the rock is only part of the equation, though. By virtue of their weather-transforming abilities, mountains attract precipitation in the form of rain and snow. Although rain is a more economical form of precipitation—one inch equals 10 inches of snow—the ability of a mountain to store water in frozen form throughout the winter months is essential for lowland life throughout the seasons. In this respect, a mountain is a sort of savings bank whose interest payment comes in the form of a good harvest. "May Kabul be without gold rather than without snow," goes the Afghani proverb, speaking to the heart of the matter. If the Hindu Kush in that country were suddenly to go snowless, that city would die—as would Denver without the snow of the Rockies, New York without the snow of the Adirondacks, Milan without the snow of the Alps, and so forth.

Yet snowless years, or at least years with decreasing snowfalls, are becoming more common around the world, and scientists point to global warming as the likely cause. The Rocky Mountains region in particular has suffered from a long drought, and the loss of glacial fields on mountains such as Kilimanjaro and Mont Blanc, among other global indicators, suggests that a major climatological shift may be at play. Even the snowpack on Everest, the highest mountain in the world, has been declining. Scientists have therefore proposed adding it to the World Heritage Danger list in 2004, arguing that the disruption of regional water cycles in that region could prove catastrophic.

No one is quite certain how to keep the well from running dry, but the water towers of stone stand ready for the happy day that the rain and snow fall in abundance again. And they will remain, as always, the proud water bearers of earth, scattering their harvest across its lands.

—Gregory McNamee

MONSOONS: MYTHS AND METEOROLOGY

A monsoon is simply defined as a seasonal shift in the prevailing wind direction, and is derived from the Arabic word for season, or *mausim*. For six months, the winter monsoon winds blow from the northeast, or offshore. This season is generally referred to as the dry season. During the summer monsoon or wet season, the prevailing winds are reversed and blow from the southwest, or onshore. This onshore flow brings moist oceanic air over the land, where it is forced to rise and create clouds and precipitation. The most intense monsoons are observed near the Arabian Sea, and consequently affect portions of southern Asia and India. It is here that some of the highest average rainfalls in the world are observed.

Monsoons are not restricted to India and the surrounding Asian countries. Monsoon circulations are also observed in Europe, northern Australia, southern Africa, Chile, and even in the United States. The Desert Southwest experiences a dry and rainy season, attributable to a monsoon circulation off the Gulfs of Mexico and California. During the late summer months, moist tropical air is transported northward over the high desert terrain of northwest Mexico, Arizona, and New Mexico. This moist air acts as fuel for afternoon and evening thunderstorms over the mountains, and at night, these thunderstorms occasionally move down to the lower-elevation desert valleys and produce some severe weather and flooding rains.

In the regions where the monsoon circulation is especially well developed, seasonal precipitation is closely linked to the onshore and offshore flow. The summer monsoon, flowing onshore, brings moist oceanic air to the land. The air is forced to rise over the land, producing cloudiness and precipitation.

—*Gregory J. Stumpf*

Hail larger than softballs and winds estimated at 80 miles per hour rendered the tornadoes spawned by this powerful gust front insignificant by comparison. The supercell dropped many small tornadoes to the east of Turkey, Texas, causing little damage. Hail and wind from this storm, however, caused widespread damage in rural areas between Silverton and Childress, Texas. The orange coloration seen in the center background originated from the sun setting behind the storm. Although the orange-colored area appears to be rain-free, very large hail was falling within the area, making it a most dangerous place to be. Photo: Alan Moller

MAKING IT RAIN

Cloud seeding, making rain by scientific means, goes back just over 50 years, but making rain by unscientific means dates from antiquity. Just about every culture based on agriculture, especially those in arid regions or areas prone to drought, had some type of system aimed at influencing rain. These were typically based on the local religion and involved prayers, offerings, animal or human sacrifices, and other rituals. Modern rainmaking may be more successful than ancient methods, but its future is uncertain.

"Rainmaking" through hocus-pocus is an age-old scam. The late actor Burt Lancaster played Bill Starbuck in the 1956 movie *Rainmaker* (based on N. Richard Nash's stage play), which chronicled the con Starbuck perpetrated on drought-weary Three Point, Texas. The story of traveling rainmakers, though, goes back through centuries. Some were lucky enough to have retired in plenty; others, depending on the times and the circumstances, faced fates ranging from banishment, to prison, to death.

Although it's possible that some of these hucksters may have stumbled onto a practical method for inducing rainfall, the concept of making rain by scientific means is only about a half-century old. In the late 1940s, a General Electric experiment in a completely unrelated field led to the chance discovery that frozen carbon dioxide (dry ice) could turn supercooled water vapor into ice crystals. This phenomenon was isolated, and the first commercial applications began about 1950.

Moments before sundown on the Lamar River in the northeastern quadrant of Yellowstone National Park in Wyoming, a breathtaking rainbow appears. Within seconds of capturing the moment, a microburst knocked the photographer off his feet and sent his 10-pound tripod flying 20 feet through the air as the storm continued. Photo: Carter E. Gowl

These days, most rainmaking activity involves airplanes, but ground-based sprays can also be effective in some cases. In practice, a variety of different sprays are used, with silver iodide crystals being the most common. Applications are done from a variety of altitudes, depending on the specific nature of the cloud that is the target.

The effectiveness of cloud seeding, measured in laboratory settings as an increase in expected precipitation, range from 5 percent to 20–30 percent, with certain circumstances yielding increases of as much as 100 percent. In general, increases are heavier in summer months than in winter, when the technique is used to increase snowpacks for water supplies or skiing. Although an increase in rainfall is the most common use for the technology, it is also sometimes used to decrease potential hail damage and to reduce visibility-hampering fog.

The silver iodide used in most seeding activity is not believed to pose a health risk; silver and iodine, the two compounds produced by the process, are diluted far below any known risk threshold. Ice crystals can also be formed during seeding with dry ice, liquid propane, or compressed carbon dioxide. In warmer conditions, the technique may instead involve the use of compounds that attract water—urea, ammonium nitrate, and a variety of salts are the agents of choice in these conditions. For cloud seeding to work, other atmospheric and meteorological conditions must be favorable. That is, there should be enough existing moisture in the air at the intended altitude and the temperature must be within a certain range. These conditions, though, might be enough to produce rain on their own, so it is often difficult to measure or verify the benefits. To cloud-seeding clients, however, the potential rewards are clearly thought to be worth the expense.

In Grant County, Oklahoma, near the Oklahoma-Kansas state line, a cold air front pushes south, running into a warm, moist unstable atmosphere resulting in a violent clash of air masses. The warm air from the south rose over the cold air, producing this beautiful shelf cloud as well as a dangerous storm. Golf ball-sized hail and hurricane-force winds of 77 miles per hour demolished a mobile home and knocked down trees and power lines. Lightning also struck a home and two oil-storage tank units, setting the home on fire and completely destroying the fire tanks. Photo: Hank Baker

A commonsense interpretation of cloud seeding suggests that the release of rain in a target area should be expected to "siphon off" rain from downwind areas. That is, if there is a given amount of moisture in the air, there should be less of it after a rain event. But those who seed clouds for a living point out that it is normal for areas downwind of the target area to also receive more rain. Depending on the geography, this is because the moisture content is constantly being replenished through evaporation from the ground.

Recently there has been increasing alarm, mostly from non-scientists, over what is called "contrail weather modification." The basic theory behind this concept is the known tendency for ice crystals to form more easily around existing ice crystals (as they do around dust or silver iodide crystals) than in clear air. With the increasing use of high-altitude aircraft and their resultant contrails, especially in high-volume commercial airways, this theory proposes that the contrails they produce increase the formation of ice crystals and clouds, changing regional weather patterns. After the terrorist attacks on September 11, 2001, all flights over the continental United States were grounded, providing a short test of the concept. The results, however, where not conclusive and both groups, those who subscribe to the theory and those who seek to debunk it, claimed victory.

Today's known and economically feasible technologies for cloud seeding may be widely used, but they are not devastatingly effective. Cloud seeding may work, but it doesn't work well enough to be a problem or a definitive solution. That may change, however, with breakthroughs in delivery or with another serendipitous technological discovery that improves reliability and efficiency. Tomorrow, the world may have to deal with the frightening possibility that freshwater could be literally plucked from the clouds before those clouds are ready to release it, changing global weather, crop distribution, and population concentrations.

—Tim Kern

RAINMAKING

Typical rain events release 10–15 percent of the existing moisture in the local atmosphere. An increase in precipitation through cloud seeding bumps this amount relatively little, to only as much as 20 percent of existing moisture.

1. The seeding compound acts on existing supercooled water droplets, causing them to freeze and initiating a chain of crystal-forming events that result in rainfall or snowfall.

2. The seeding activity and the formation of ice crystals release latent heat from the fusion process itself, raising the temperature of the surrounding cloud, thereby producing a larger cloud and a larger precipitation event.

FUTURE PROBLEMS

Cloud seeding, as the state of the art exists today, generates only small changes in local weather. Future gains in effectiveness, though, could raise moral and legal questions as the amount of water produced increases.

- Who owns the water in the clouds over a given area? Is it a person or a state?
- If water is pulled from clouds upwind and the region downwind suffers, what would be the legal liability on those who seeded the clouds?
- Thinking not only of the long-term effects of triggering drought downwind (and its consequences, such as forest fires and famine), what could happen in the "target" area if the seeding were too successful?
- Could widespread cloud seeding, over time, cause significant, near-permanent climatic damage?
- There's always "Mom's Rule," as well. All moms know it, in one form or another; it's a good rule to consider before making any major decision: "What if everybody did it? Would everybody be better off?"

WEATHER WOES DOWN ON THE FARM

LEFT: *A snow shower falls from "out of the blue," over a Lincoln County, Wisconsin farm. Radiant heat from the sun produced scattered clouds, some of which produced rain. But as the afternoon sun waned, unusually cool temperatures changed the rain to snow showers, producing the curious sight of an isolated cloud showering snow against a blue sky. For about a half hour, only a few of the clouds developed snow, as the clouds quickly dissipated when the temperatures dropped. Photo: David L. Sladky*

ABOVE: *Examples of "super-sized" hail displayed by some astonished residents. Photo: Courtesy of NCAR*

Hail threatens agricultural production worldwide. Once thought to be an unfavorable sign from the gods, it wreaks havoc from those Texas Panhandle fields and the Midwest all the way to China, sometimes leaving half a field in tatters and the other half untouched. In the United States, we often hear about the destruction of two big crops, corn and wheat, but hail also destroys citrus and other fruits, cotton, tobacco, tea, and vegetables. Although it's easy to believe that the bigger the hailstones are, the harder they fall, when it comes to crops, small hailstones—since there are more of them—can be more destructive than large ones, especially to tender young plants. One researcher estimates annual U.S. crop losses due to hail at $1.3 billion, a figure that represents 1 to 2 percent of the total annual crop value.

Hailstorms may deliver the wrong kind of precipitation, but it's the lack of precipitation that farmers fear most. The deadly killer down on the farm is drought. The official government definition for drought is "a period of insufficient rainfall for normal plant growth, which begins when soil moisture is so diminished that vegetation roots cannot absorb enough water to replace that lost by transpiration." Crop scientists also use other definitions based on meteorology, hydrology, and socioeconomics. And in some parts of the country, drought can also be linked to winter precipitation, because the snow that falls in the winter in the mountains is relied upon for irrigation in spring and summer months. Too little snow means not enough water for irrigation.

Over a period of a few years or more, a continuing drought can lead to a "Dust Bowl" stage in which the moisture content in the soil is seriously depleted, leading to a permanent loss of topsoil as it is blown away. Ranches also are hard hit by drought; water shortages reduce the amount of livestock that can be raised and bump up the cost of feed, often forcing the sale of animals for below-market value. Though measured in billions, the losses caused by drought to farming are hard to measure in simple dollars, because drought has such far-reaching effects in the social and economic fabric of a given region.

In crop losses for a single commodity (wheat), the recent drought in Colorado resulted in an estimated $100 million for the 2002 season, representing about half the value expected if precipitation were normal. Farmers can rely on "multiple-peril insurance"—only about 30 percent of farmers carried such insurance as of the mid-1990s—or government assistance in the form of low-interest loans, but during a drought cycle it's easy to become "loaned out."

Drought teaches lots of lessons: new conservation methods for soil and water; new farming techniques, such as crop rotation; and the development of more drought-resistant varieties of wheat and other crops. Precision farming applications, still experimental, use satellite-positioning technology to measure growth patterns and soil moisture. Sensors in equipment can regulate the amount of seed, fertilizer, and water for irrigation, varying these amounts if needed, from field to field or even within single fields, as recorded measurements dictate.

From too little to too much, flooding causes farmers all the serious problems regular homeowners face, and then some. The Great Flood of 1993 on the Mississippi River, which lasted from June through August and raged through 10 states, offers the best (or worst) recent example of farm woes from flooding. Total damage estimates range from $12 to 21 billion, but it's hard to identify the portion represented by agriculture, since no single government agency is responsible for systematically collecting or reporting flood-loss information. Service agencies and organizations that provide assistance also reveal problems faced on farms during a flood. There's more than just rotting crops, mold, fungus, and mildew to deal with; negative effects can also include the dilution of manure spreading into unwanted areas—including drinking water supplies, washed out feedlots, the saturation of stored grain, and water-damaged hay bales. And if the latter isn't damaging enough, spontaneous combustion can also occur from soaked hay overheating as it dries. After floods, danger lingers. Some fields may produce lower crop yields because of the altered chemical nature of saturated soil; seeds carried in by floodwaters can include those of new and different weeds; plant diseases linked to soil moisture increase; and tractors working sodden fields can over-compact the soil.

And then there are the rats and the mosquitoes. As the Mississippi floodwaters rose in 1993, an Iowa State University Extension wildlife specialist advised farmers to poison rats to prevent the spread of disease. Fields full of standing water create breeding grounds for mosquitoes and other disease-carrying insects that threaten livestock as well as humans. And if the thought of having to clean up the chainsaw after the basement's flooded makes you cringe, imagine what it would be like to recondition a sodden tractor or combine.

With the benefit of weather forecasting and flood warnings, today's farmers may be able to dodge more of Mother Nature's extremes than their predecessors. But even though tractors or livestock can be moved to safety, wheat, corn, and soybeans have no choice but to remain out in the open, vulnerable to weather woes, just as they have for thousands of years.

—Kathleen Cain

RIGHT: *A "high precipitation" (HP) supercell that occurred in the Texas Panhandle has considerable precipitation under its updraft base (left side of image) caused by light upper-level winds. The rain in these supercells falls closer to the updraft and is then pulled around the storm's base by the swirling mesocyclone, resulting in a menacing-looking rain-wrapped mesocyclone. This storm produced large hail, damaging winds, and several tornadoes that were difficult to see because of the dark precipitation. Photo: Alan Moller*

FOLLOWING PAGES: *The luck of the Irish was in Otjiwarongo, Namibia, Africa, when these lenticular clouds over the Omataco Mountains were photographed on St. Patrick's Day, 2004. These clouds were formed in a moist layer of stably stratified air that was forced to ascend when the wind flowed over the mountain. The rising air then cooled and formed clouds over the mountains when the temperature fell to the dew point. On the downwind side of the mountain, descending air forced the clouds to dissipate. Stable stratification suppressed vertical mixing of the air, so pronounced variations in moisture content occurred in narrowly separated layers. Photo: Viveca Venegas*

Clouds change shape as parts evaporate in dryer air or when they are moved and reshaped by wind and air currents.

LUKE HOWARD:
THE MAN WHO NAMED THE CLOUDS

For millennia, clouds passed through the skies as interesting but nameless shapes. They were observed, painted, rhapsodized over in poems, books, plays, and sonnets. They were compared to animals and angels and described in terms of their form and color. But they had no names. Until an evening in December of 1802. That night, in London, a small society called the Askesians held one of its regular meetings. Luke Howard, a young pharmacist who had been fascinated by meteorology since childhood, read a paper to his fellow members in the society. His paper was entitled "On the Modifications of Clouds." Little did any of the people in this small group know the lasting and momentous effect this paper would have on the world of meteorology. In the paper, Howard gave Latin names to the most commonly seen clouds: cumulus, stratus, cirrus, and nimbus. Those names, while amended and added to over the years, still form the basis of cloud classification today.

Why did the naming of clouds occur so late in history? After all, clouds have been influencing human behavior and activities for eons, bringing rain, blocking sun, simply being there to be seen. The answer no doubt has to do with their ephemeral nature. Clouds appear and disappear; they change shape, move, or hang still. How could such amorphous objects be analyzed and classified with any reliability? How, in fact, could scientists be expected to take these whimsical forms very seriously at all? Scientists look for order in the universe. The regular movements of heavenly bodies—stars, planets, the sun, and the moon—these were all worthy of scientific study. But a study of the capricious movements of clouds and storm systems? Not likely.

Unique to the atmosphere of Australia, a Morning Glory cloud advances over the mangrove salt marshes of the Gulf of Carpentaria. These clouds may be up to 620 miles long and have a sculpted, undulating shape along their entire length. These giant clouds form seasonally when specific climate conditions interact with the unique geography of the region, most commonly when opposing sea breezes collide over the hot landmass of Cape York. When there is sufficient humidity, the cloud forms as humid air is forced up the front of the wave and condenses. On the descending edge of the wave, the cloud evaporates, creating the awesome effect of the entire cloud mass revolving backward, while advancing at 37 miles per hour. Photo: Barry Slade

Given all that, it's understandable why astronomy preceded the development of meteorology by centuries. Yet in time, scientists' attention did turn to the weather around them, and to the clouds above them. As with any science, a number of inventions were essential before a systematic study could be undertaken. For meteorology, those inventions were the barometer and thermometer, which were in general use throughout the 18th century. During that century, many people with a curious nature began keeping weather diaries. By the end of the century, the laws governing the behavior of gases were becoming reasonably well understood, as was the nature of the sheath of air that surrounds the globe.

Scientists began to understand that the atmosphere did obey the laws of physics and chemistry, and that its future state could be predicted if its present and past states were accurately known. This implied careful measurement of the atmospheric variables of pressure, temperature, moisture, and wind, both at the surface and in the upper air. But for a true ability to predict weather and weather patterns, measurements would have to be made at many places over the globe, all at the exact same moment in time. For that to happen, meteorology had to wait until the latter part of the century and the invention of the telegraph to make precise timing possible.

Luke Howard was born in London on November 28, 1772. His father was a staunch Quaker, and young Luke attended a Friends grammar school near Oxford. It was here that he learned Latin, more Latin there than he would ever be able to forget, he commented later. His proficiency in the subject would play a tremendous role years later when he developed his classifications.

There is no doubt that the phenomenal weather year of 1783 was an important part of Howard's early fascination with clouds. In that year the young Howard, already a devoted observer of the universe, saw nature in a state of extreme agitation and turmoil. In May and June, violent volcanic eruptions shook

Iceland. The activity continued through the year, sending forth the greatest lava flow known to man. Great volumes of volcanic dust from the Eldeyjar eruption fell over Iceland and, carried by the westerly upper-level wind streams, spread over Scotland, destroying crops and killing livestock. The dust cloud eventually spread across England, continental Europe, and reached all the way to North Africa.

Meanwhile, another eruption was occurring on the other side of the world. In August, a volcano in Japan called Asamayama ejected boulders as large as houses and spewed enormous quantities of dust into the upper atmosphere. Westerly winds carried the dust around the whole Northern Hemisphere, which added to the general pall of haze. All the weather diaries of that period and even some of the great literary works of the day contain vivid references to the extended period of the "Great Fogg". The sense of underlying uneasiness was heightened by major earth tremors in Calabria and Sicily.

Finally, and most gloriously, on August 18 a large fiery meteor flashed across the skies of Western Europe. It was seen by tens of thousands of people, including young Luke Howard. There can be little doubt of the effect these events had on a 10-year-old boy who was already cloud-struck.

With his school days behind him, Howard became apprenticed to a chemist in Stockport. He endured years of drudgery as he ground chemical preparations, cleaned bottles, and swept floors. He satisfied his intellectual cravings after hours, by studying French, botany, and chemistry. Eventually he became a pharmacist and opened his own business in London. Soon he became friends with other young men with similar interests in scientific matters. One particularly close friend was William Allen, who owned a successful commercial pharmacy. He hired Howard to run his manufacturing laboratory in Plaistow, just outside London. As Howard traveled back and forth between London and Plaistow, he had ample time to continue his studies of cloud forms.

It was Allen who decided, in March of 1796, to establish a society for the discussion of scientific matters. This society was dubbed the Askesians, from the Greek *askesis*, which means, roughly, intellectual exercise. The meetings were held every other week at 6:00 p.m., and the rules required that each member prepare a paper on a subject of interest, and read it before the society, in rotation, or pay a fine.

While Howard was thoroughly engaged with the Askesians, he was also doing a great deal of scientific study on his own. He was particularly impressed with the work of Swedish taxonomist Carl von Linné, known more familiarly as Linnaeus. Linnaeus had established the beginnings of the modern system of classification of all life forms, a system scientists around the world would eventually adopt.

Howard presented a number of papers during the six years that he had been a member of the Askesians, but it was the paper that he presented on an evening in December 1802 that ensured his reputation. His lecture, "On the Modifications of Clouds," was a brilliant exposition of a scheme he had developed for classifying and naming clouds. It was the culmination of Howard's years of observation. Two centuries later, we are still using the essence of his scheme.

Howard proposed in his paper that it was possible to identify a number of simple categories within the complexity of the changing skies. He set out to establish a complete classification that would cover all possible cases. Like Linnaeus, he used Latin names, a language that fortuitously transcends national boundaries. He put clouds in four groups:

Cumulus (Latin for heap) Convex or conical heaps, increasing upward from a horizontal base
Stratus (Latin for layer) Widely extended horizontal sheets
Cirrus (Latin for curl of hair) Fibers that can stretch in any or all directions
Nimbus (Latin for rain) Systems of clouds from which rain falls

One of the principal insights in Howard's essay is his insistence that clouds are a proper subject for research and theory. He noted that clouds have many shapes but only a few basic forms. Those shapes and forms are caused by water present in the atmosphere, water that may be in any or all of its three states: liquid droplets, solid ice crystals, or gas (water vapor). He noted, too, that while circulating air causes unstable conditions in the atmosphere, the principles of cloud formation are still understandable and are capable of being classified and studied scientifically. Howard's other important assertion was that clouds can change from one form to another. A cumulus cloud may flatten and widen into a stratus formation. Simple observation shows that clouds move, but Howard contributed the knowledge that they change and merge. They are not fixed in one form for easy classification and study.

Even though the physics of air and water vapor were poorly understood in Howard's day, his analysis had a sound physical basis, and his paper was met with acclaim. It was soon published in Alexander Tilloch's *Philosophical Magazine,* a major disseminator of scientific knowledge. Before long Howard's paper was reprinted in a number of various journals and encyclopedias, both in England and on the Continent.

Howard continued his studies of meteorology, and his work included many significant achievements. In 1806, he began his Meteorological Register, which was published regularly by *Athenaeum Magazine.* He wrote the first book on urban climatology, *The Climate of London,* published in two volumes in 1818–19. In it he wrote of the impact that cities can have on meteorological conditions. For example, he carefully described London's famous fog, and noted that there could be almost no visibility in the city while a few miles away the atmosphere was clear. This seems very obvious today and not particularly noteworthy, but smog, as we call city fog today, was an unknown concept in the early 19th century.

In 1821, Howard was elected a Fellow of the prestigious Royal Society, for his contributions to meteorology. In the second half of Howard's life, he became more involved with his chemical manufacturing business—though his interest in clouds never flagged. His long life ended on March 21, 1864. At his funeral, his son said, "A beautiful sunset was a real and intense delight to him; he would stand at the window, change his position, go out of doors and watch it to the last lingering ray." Clouds in all their many moods carried an endless fascination for the man who invented their names.

Eventually, a list of 10 cloud types was drawn up, expanding on Howard's four.

Cumulus, Stratus, Cirrus, Nimbus, Cirrostratus, Cirrocumulus, Stratocirrus, Cumulocirrus, Stratocumulus, Cumulonimbus

With minor revisions, these appeared in the 1896 *International Cloud Atlas* and are still in use today.

Luke Howard was untrained in meteorology, a chemist and pharmacist by profession, yet through his never-ending curiosity and masterful powers of observation, he left an indelible mark on the world around him. Some call him, with good reason, the Godfather of Clouds.

—John A. Day

TEN REASONS TO LOOK UP

1. Clouds and cloudscapes are the greatest free show on earth. It doesn't cost a penny to look up and feast your eyes on the view.

2. Clouds are never exactly the same. While there are four basic cloud types (cumulus, stratus, cirrus, and nimbus), nature combines them to compose endless symphonies in the skies.

3. Many skies are simply beautiful to behold. There is no other way of saying it. The gradations of light and color in the late afternoon and very early morning hours are bouquets for the eye.

4. Clouds are a billboard of Coming Attractions. While it takes a skilled eye to interpret the messages on the billboard, there is a feeling of immense satisfaction when one's own forecast is verified.

5. Observing the sky at regular intervals makes one feel connected to nature.

6. Cloud watching promotes a global consciousness. Weather satellites bring large-scale images of cloud patterns into our homes. They help us realize that "our" clouds are connected to other clouds all around the world.

7. The earth is unique because of its vast amounts of water. Clouds are made of water and are a constant reminder of its importance.

8. Water is a miracle substance. Scientists have found that simple H_2O is anything but simple. Those H_2O molecules link together and bring us the glorious clouds above us. Without water, there would be no clouds.

9. Cloud watching is an antidote to boredom. Clouds are ever changing, ever evocative.

10. Clouds are a magic show. Where do they come from, and where do they go? This is a mystery to the nonscientist, and an area of endless fascination.

— John A. Day

UNUSUAL CLOUD TYPES

1. **VIRGA:** As rain clouds gather, if you see a tail or curtain or veil of cloud flowing beneath them, or the rain looks as if it is walking above the horizon, it may be the virga slipping away through the sky. Although it can develop any place where precipitation occurs, it is most likely to develop in drier climates like Denver, Colorado. The nearby mountain barrier robs the low-level atmosphere of moisture, making the air even drier beneath the clouds. When the moisture leaves the upper atmosphere and descends in the form of the serpentine virga, its chances of evaporating before it reaches the ground are much higher.

2. **MAMMATUS:** They are often seen near the back edge of strong to severe thunderstorms. They are farmed by packets of "reverse" convection, i.e., the downward acceleration of air parcels. These pockets create the downward protrusions within the thunderstorm clouds.

3. **IRIDESCENT:** The iridescence is seen as a patchwork of pastel color throughout the clouds, and is produced by the diffraction of sunlight traveling through various-sized cloud droplets.

4. **KELVIN-HELMHOLTZ WAVES:** These delicate wave clouds are also referred to as billows. They occur when the wind velocity changes rapidly with height, resulting in breaking waves, similar to what occurs at the beach. When enough moisture is present within the rising-motion portion of these waves, then condensation occurs, thus allowing us to see these waves. In dry conditions, the waves are still present but occur in clear air. Kelvin-Helmholtz waves are the principal cause of clear-air turbulence for aircraft.

5. **CUMULONIMBUS:** The cauliflower-like portion of this thunderstorm cloud beneath the flat anvil represents the region of the storm where air is rapidly ascending in an unstable atmosphere. Here, the cloud is composed of supercooled water droplets, or water in its liquid state in a subfreezing environment. The flat ice-crystal anvil cloud forms at the base of the stable stratosphere, often eight miles high or higher, where the cauliflower updrafts from beneath lose their buoyancy and spread horizontally.

6. **LENTICULAR:** A very smooth, round or oval, lens-shaped cloud that is often seen, singly or stacked in groups, near or in the lee of a mountain ridge. These clouds are caused by a wave wind pattern created by the mountains. They are also indicative of downstream turbulence on the leeward side of a barrier.

—Kathleen Cain

LEFT: *South of Beloit, Kansas, a contrail billows in the evening.*
Photo: Charles A. Doswell III

VIRGA *Photo: Jeff Foott/Getty Images*

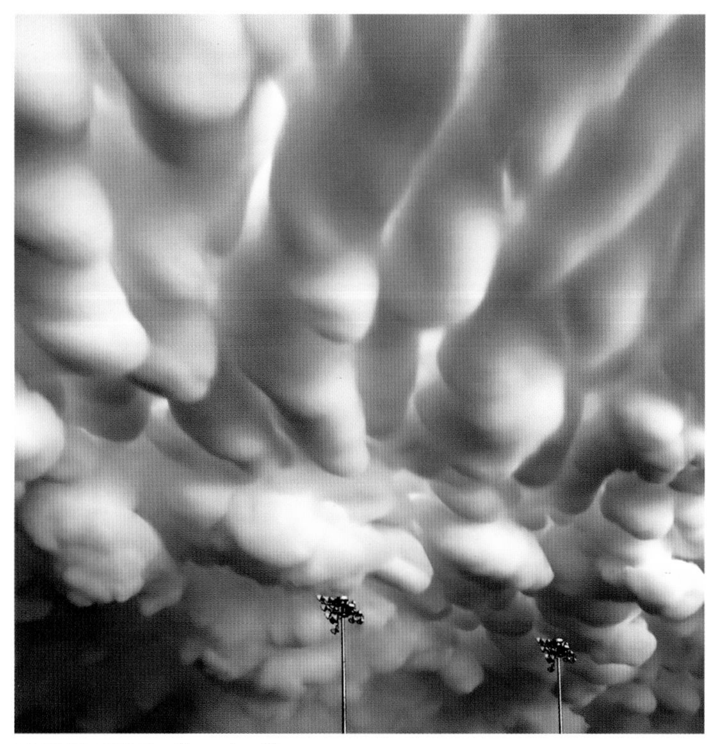

MAMMATUS *Photo: Jorn Olsen*

IRIDESCENT CIRRUS *Photo: Warren Faidley/Weatherstock*

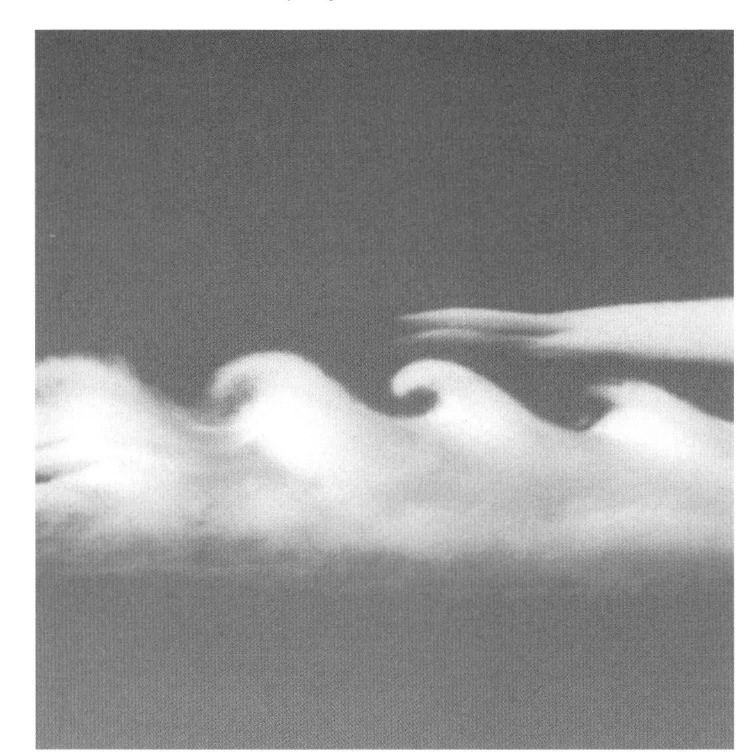

KELVIN-HELMHOLTZ WAVES *Photo: Brooks Martner*

CUMULONIMBUS *Photo: Ricahrd Kaylin/Getty Images*

LENITCULAR CAP CLOUD *Photo: Stan Osolinski/Dembinsky Photo Associates*

The coast of California is one of the foggiest locations in the U.S., with more than 60 days of dense fog reported on average per year.

COASTAL FOG:
BEAUTIFUL AND DEADLY

PRECEDING PAGE 158: *A classic lenticular cloud is captured in the Sierra Nevada outside Reno just minutes before the onset of a windstorm so severe that it cracked the windows and stripped the paint off the photographer's car. Photo: Kathleen Norris Cook*

PRECEDING PAGE 159: *The hanging, pillow-like pouches underneath the anvil top of certain cumulonimbus clouds (mammatus clouds) like these in Weld County, Colorado, indicate severe turbulence. Photo: Gregory Thompson*

LEFT: *A ridge of Claremont Canyon rises over the fog-draped San Francisco Bay. Photo: Galen Rowell/Mountain Light*

ABOVE: *A pillow of fog rests below Mt. Edgecumbe, in Sitka Sound, Alaska. Photo: Fred Hirschmann*

It is difficult to understand how something that inspires so much romance can also create so much havoc. No wonder that the word fog also means an uncertain or confused position.

Fog is formed by the condensation of water vapor on condensation nuclei, microscopic dust particles that are always present in natural air. This condensed water vapor in cloudlike masses lying close to the ground is sufficiently dense to reduce horizontal visibility to less than 3,281 feet (1,000 meters). As soon as the humidity of the air exceeds saturation by a fraction of 1 percent, fog occurs. Coastal fog, like that common in San Francisco, is technically known as "advection fog." As the warm, moist air moves over cold surfaces, it is cooled below the dew point, the temperature at which dew forms. Fog will occur either when the moisture content of the air is increased beyond the saturation point or when the air is cooled below the dew point, causing the excess moisture to condense and form fog.

The advection fog that is common along the Pacific Coast· in summer comes from warm, moist air that is carried by the wind from warmer waters off shore. When the warm air reaches the surface waters near the coast, the air cools and condenses due to a seasonal upwelling this region of the ocean experiences, which pushes colder water to the surface. After the sun has had a chance to penetrate the fog and warm the earth, the conditions are altered, and the fog disappears.

The two major coastal fog areas in the United States are the Pacific Coast region and the New England region. San Francisco does get lots of fog—more than 18 days of solid, heavy fog are recorded on an average each year—but San Francisco is not the foggiest place on the West Coast. Cape Disappointment, Washington, located at the mouth of the Columbia River, earns this title. Here, the average is 2,552 hours of heavy fog a year. That translates to nearly 106 days of fog. Who do you think had the honor of naming this community? A disgruntled fisherman, no doubt.

On the East Coast, the foggiest area is along the Maine coastline. Moose Peak Lighthouse, altitude 72 feet, is situated on Mistake Island, where they average 1,580 hours of heavy fog a year. No wonder the waters

BELOW: *The Grand Tetons in fog on a fall day. Photo: Tom Vezo*

RIGHT: *Fogbows are relatively common in moisture-laden regions but are quite rare in desert environments, such as Turret Arch in Arches National Park, Utah. White fogbows are a close cousin of colorful rainbows. The cloud droplets that form fogbows, however, are much smaller than typical raindrops. The light-ray paths through these tiny droplets are too short to allow for the spectral separation of the white light, which is why fogbows are typically colorless. Photo: Paul Neiman*

off the coast of Maine rank as some of the deadliest in the world for fishermen. Despite its beauty, fog can kill. A Pacific steamer named *City of Rio de Janeiro* struck a rock while entering San Francisco Bay in dense fog. The ship sank in a matter of minutes, and 128 Japanese and Chinese drowned. The Italian luxury liner *Andrea Doria* was rammed by the Swedish liner Stockholm on July 25, 1956. The *Doria* was approaching the East Coast, completely obscured by fog at the time. The two liners saw one another on radar, but there was tremendous confusion, and they rammed together. Fifty people lost their lives.

Fog can also support life, including water-hungry plants that are found along the California coastline. In Chile, fog provides water for people. Twenty years ago, the only water for Chungungo, a small village on the north-central coast of Chile, had to be trucked in from more than 30 miles away. Today, this same village receives their entire water supply from coastal fog. A ridge above the village is frequently blanketed by advection fog, making it ideal for such a project. Fifty large fog collectors, made of polypropylene mesh, collect condensed water from fog, route it into a large storage tank, and then into village homes. Each of the 330 villagers is allowed 33 litres of water per day! (The average American uses about 10 times that amount.) And even though the fog water may contain dust particles and a little salt, it meets Chilean drinking standards.

The best way to predict fog is to know the wind direction and the surface temperatures along the shores, over both water and land. Also, it can be predicted by obtaining the relative humidity (a measure of how saturated the air is) from an observation station upwind from you. If the air mass is nearly 100 percent humidity, you can expect fog.

Suddenly encountering fog while driving on a highway can be a frightening and horrifying experience. The best way to drive through fog is with your low beams on. The high beams reflect off moisture particles in the fog and bounce light directly back into your sight. Better yet are fog lights, installed by the manufacturers on many newer model cars. They illuminate the area directly above the surface of the road and below the fog bank, allowing maximum visibility in a heavy fog.

In regions where coastal fog hampers movement on the water, other methods have long been used to help guide boats. A fog signal is a sound or light signal sent out by lighthouses and buoys to indicate a shoreline, a channel, or a dangerous stretch of water, and by vessels to mark their positions. Each signal has a distinctive code. All vessels, moored or stationary, are required by law to use fog signals in bad weather; the type, number, length, and timing of the signal indicate the size of the vessel and its position. The earliest sound signals were made by bells, gongs, and even explosives. All have been replaced by foghorns.

—*Kate James*

WEATHER AND THE WORLD'S TALLEST TREES

Sempervirens, from the Latin words meaning "evergreen," scientifically identifies one of the planet's most celebrated life forms: the coast redwood. Trees of the species *Sequoia sempervirens* soar in height, the tallest of them rising to more than 350 feet above their forest floors. Although coast redwoods may be high reaching, abundant they aren't. The redwood's geographic distribution resembles its physical stature: long and slender, spanning roughly 470 miles of rugged Pacific coastal mountains—and that's it. Never more than a few miles inland of the cold, crashing surf of the Pacific, the redwood's narrow habitat is anchored on its southern extremity by a small grove in the Santa Lucia Mountains of Monterey County. Redwoods of the Chetco River in Oregon, about 15 miles north of the California-Oregon border, stand as the species' northernmost members.

The coast redwood isn't just tall. It is the tallest living organism in the world. As of 2004, the Mount Everest of redwoods, the Stratosphere Giant, stood 370 feet, 2.4 inches above its base—8.4 inches higher than it was when it was discovered in 2000. Located in the Rockefeller Forest of the Humboldt Redwoods State Park near the town of Redcrest, California, the Stratosphere Giant just barely eclipses (by 5.1 inches) the planet's second tallest life form, the National Geographic Tree—also a coast redwood. In fact, 15 coast redwoods in total stand above 360 feet. The heights of these trees pale in comparison to the theoretical maximum loft of a Sequoia semperviren, however, which scientists have established to be 425 feet! But without the climate unique to California's northern coast, these trees wouldn't grow at all.

Cool, wet winters and warm, dry summers describe the sempervirens' ideal climate—almost. The Pacific Ocean moderates the temperature of the redwood's habitat year-round; Eureka, California, which lies in the heart of redwood country, is renowned as having one of the most stable annual temperature regimes of any place in the world. But the ocean also generates soaking storms that drive into the northern California coast throughout the winter months. Precipitation falls heaviest on the northern reaches of the redwood's habitat, where the sempervirens stand tallest. Redwoods of the southern portion, while robust, rise to heights lower than those of their more northerly kin. Average annual precipitation varies from about 20 inches in the south to 120 inches in the very wet, northern reaches of the trees' habitat.

Fog-loving coast redwoods in Muir Woods National Monument, Marin County,California. Photo: Galen Rowell/Mountain Light

BELOW: *Sea fog creeps up the cliffs of St. George Island on the Bering Sea in Alaska. Photo: Fred Hirschmann*

RIGHT: *Sky-high redwoods like these in Humbolt County, California, near the town of Redcrest, are some of the tallest living organisms in the world. Without the climate unique to California's northern coast, these trees wouldn't grow at all. Photo: Ed Darack*

Scientists who study redwoods have pondered for years why this tree grows only along this narrow strip of land. The answer lies in the occult. Not in the occult of legend and lore, but in what meteorologists call occult precipitation, as in hidden precipitation (because it doesn't register on rain gauges): fog.

During the comparatively dry summer months, two powerful meteorological forces merge along the redwood coast: moisture-laden northwesterly winds glide over cold deep-sea currents that have coursed their way to the ocean's surface. The resulting advection fog quickly runs aground—up the steep coastline and into and throughout the redwoods themselves. As the prevailing winds force the fog up the mountainous coast—a process called orographic lifting—this advection fog lifts warm air above it, which in turn cools adiabatically (temperature loss with altitude), causing moisture to condense to form another blanket of fog. And what a blanket it can be: the difference in air temperature above and below this layer can range as much as 65 degrees!

In addition to moderating summer temperatures, coastal advection fog acts as a sort of "precipitation bridge" during the summer months. As winds push these dense fog banks through the redwoods, tiny droplets collect on the trees' intricate boughs. Redwoods can absorb a small amount of this collected water directly into their vasculature, but the majority of this captured fog falls to the ground, where the tall trees' shallow roots lap it up. This "redwood rain" can be substantial: researchers recorded more than three inches (8 cm) beneath a redwood in Humboldt County during a two-day period. Some studies have even concluded that fog contributes as much as a third of a redwood's annual water input—not to mention what it contributes to other plant life and forest streams and rivers.

—*Ed Darack*

FOG FACTS

- Technically speaking, fog is present when horizontal visibility drops below .62 miles. Heavy fog is defined as visibility below one quarter of a mile.
- Fog seldom forms when the dew point temperature differs from the air temperature by more than four degrees Fahrenheit and winds are more than 10 miles per hour.
- Sea smoke is fog generated in open water from conditions that are the reverse of coastal fog—cold air and warm water. At an extreme, this ocean phenomenon can produce plumes of dense fog several hundred feet high.

Lake Effect snow occurs when very cold winds move across long expanses of warmer water. It occurs throughout the world, but is best known in the Great Lakes area of North America.

LAKE EFFECT SNOWS

LEFT: *The temperature was –27 degrees Fahrenheit and the wind chill factor approximately –80 degrees Fahrenheit when this photograph was taken along the north short of Lake Superior in northeastern Minnesota. When a cold air mass comes in contact with the warmer waters of a large lake, condensation forms clouds of steam that rise above the lake's surface, creating a tremendous amount of instability. This is the same weather dynamic that produces lake effect snow. Occasionally a "twister" of very warm air currents can form. The twister may be caused by frictional turning of the prevailing wind as it comes in contact with the lake shore. Photo: Layne Kennedy*

ABOVE: *A cold front advances over Lake Michigan. Photo: Stephen Graham/Dembinsky Photo Associates*

FOLLOWING PAGES: *The magnificent, 42-mile-long Columbia Glacier in Prince William Sound, Alaska, produces icebergs by "calving," a process in which large chunks of the ice break off into the water. In recent years, iceberg production has increased dramatically as the glacier retreats due to climatic change. Photo: Rich Reid*

Many of us have heard about the heavy snows that fall every winter downwind of the Great Lakes. When arctic air picks up moisture and heat energy from the lakes, places like Syracuse and Buffalo, New York, and Marquette, Michigan, can wind up with legendary snowfalls. Sault Ste. Marie, Michigan, received nearly five feet of snow in just a few days early in December 1995. A couple of days later, a record 38 inches in 24 hours paralyzed Buffalo as the cold southwesterly winds blew down the entire length of Lake Erie.

The type of lake effect that produces these types of snows is called an enlarged snowband. Most lake effect snows occur under lines of cumulus clouds, often called cloud streets, spaced about four kilometers apart. Since they are so tightly packed, each cloud street only has a four-kilometer-wide area of lake moisture and heat to convert to snow. Enlarged snowbands are a different animal altogether. These beasts can take in lake-induced heat and moisture up to 50 kilometers on a side and more. The result can be snowfall rates approaching the ridiculous, nine inches per hour or more.

You can view an enlarged snowband as a large heat engine, in some ways, like a hurricane. Strong converging low-level wind jets start to form well outside the snowband and converge into its central axis. Where these jets cross over water, they pick up heat and moisture and transport them to the center of the band. When the jets converge at the center they erupt vertically to create cumulonimbus clouds and heavy snow.

The analogies of an enlarged snowband to a tropical cyclone are amazing. Both create their own low-level wind flow that accelerates toward their respective centers. Both have strong updrafts and an upper-level exhaust system (or outflow) that carries the excess heat away. The analogy even goes further because the center of an enlarged snowband is an area where the winds die down, much like an eye of a hurricane. The analogies end there because a snowband is a linear feature whereas a tropical cyclone is circular. And because an enlarged snowband is a wintertime feature, there is not enough heat and moisture to create the damaging winds or the deep cumulonimbus clouds. Even though enlarged snowbands are impressive in the lake effect world, they rarely exceed 15,000 feet in depth. Hurricanes exceed 40,000 feet in depth.

—Jim LaDue, NOAA

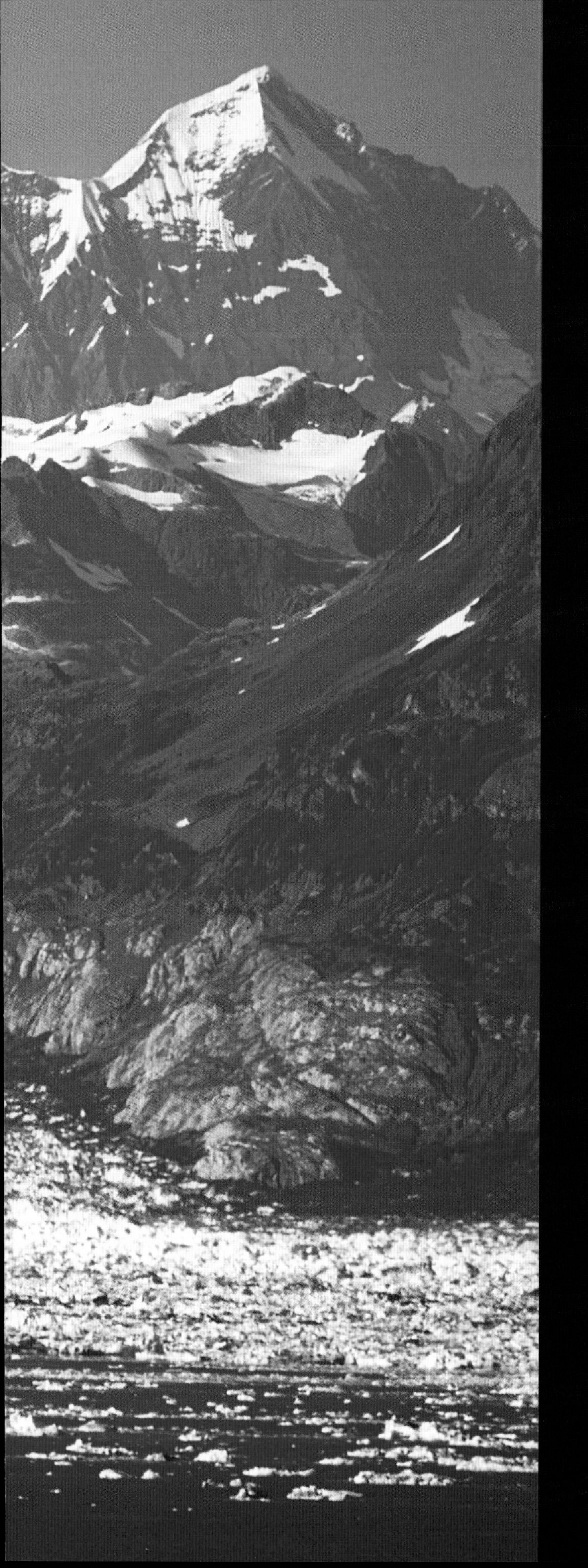

THE COMPLEX WORLD OF SNOWFLAKES

All snowflakes are not made alike, but they all start out from the same source. Snowflakes are crystals of frozen water. Sometimes they are single, simple crystals, but more often they are complex crystals or groups of many such crystals, mechanically and sometimes thermally hooked together into shapes that cannot be easily characterized.

Virtually all rain starts as snow. Higher temperatures encountered as the snow descends turn some snow crystals into drops of rain. A single raindrop can often consist of hundreds of former snow crystals. How cold does it have to be to keep snow from falling? There is no such temperature as "too cold to snow" because wherever cold temperatures are found on earth, snow falls.

Crystals have distinct patterns, or lattices. One common trait of snowflakes is that they are made of crystals that have shapes dictated by the number six: six sides, six arms, and six facets. Water makes a six-sided crystal when frozen. The final snowflake, though, may be a complex crystal, made complicated by its natural inclination to attract additional water vapor. The original structure expands in size and complexity as the water vapor freezes. A single snowflake can have up to 100 million molecules of water, and the variety of ways these molecules might form as crystals is too large to compute on a pocket calculator.

Snowflakes tend to belong to one of two broad structural classes, flat or cylindrical, largely determined by the surrounding temperature. That fact was documented in the second century BC by the Chinese, then by Albertus Magnus about 1260, and finally by Johannes Kepler in 1609. According to Kenneth Libbrecht, the chairman of the physics department at Cal Tech, flat-looking "plate" snowflakes are produced at temperatures just below freezing and also at temperatures considerably below freezing. The temperature range in between tends to produce more cylinder-shaped crystals. However, no one yet knows why certain temperatures tend to produce certain typical structures.

Humidity, too, plays a part in the snow crystal's shape. It is also instrumental in the crystal's ultimate growth. As is intuitively obvious, higher humidity tends to beget more-complex snowflakes. Less intuitive is the empirical finding that more of the cylindrical-shaped snowflakes appear in conditions of higher humidity. To study any of these factors of snowflake life, scientists have to look at them up close and in detail. The essential problem, however, is that it's hard to capture and preserve snowflakes, a necessary step in their study.

One way to capture them is to produce snowflakes in laboratories by using artificial conditions, but these specimens may not approximate nature's work. So far, the best way found to study these natural marvels is to photograph them, preserving the essence of their structure. Wilson A. Bentley, a Vermont farmer, was fascinated with just this concept, and he developed photographic methods to observe snow crystals. He is credited with being the first to photograph individual snowflakes, in 1885, an outgrowth of his early scientific quest. "Snowflake Bentley," as he came to be known, produced more than 5,000 photos of snowflakes and popularized the notion that "no two snowflakes are alike." By the time he died in 1931, this idea had become universally accepted.

But beautiful and clever as his photos were, Bentley wasn't universally acclaimed. His neighbors, he told reporter Mary Mullet, in the February 1925 edition of *The American Magazine*, "still think I'm a little cracked." Perhaps that impression was due to the awkwardness of the methodology he employed to gather the subject material, or the strain of making sure the work was timely—being outside in weather conditions that drove most people indoors. Bentley's approach to this challenge was to light the snowflakes from behind, producing a snowflake-shadow image on a light gray background. Then he hand scraped the gray background away from the snowflake's image on the negative, leaving a clear background. When he printed a negative, the clear part of the negative produced a black background against which the snowflake stood in contrast.

In the 1950s, Japanese scientist Ukichiro Nakaya applied more science to the photographic observation of snowflakes, adding his own images and observations to the sum of knowledge. Nakaya, in addition to pioneering new methods of photographing the tiny crystals, is credited with being the first to deliberately create snowflakes under laboratory conditions. In his photographs, Nakaya, like Bentley, used back-lighting, but he let the background remain, unmodified. To bring out the surface shape of the snowflake, he used a grazing sidelight that showed more of the surface texture of the flake.

Today, most snowflake photography still uses some form of the backlighting technique. Keeping snowflakes in place while they are photographed poses an equal challenge. As a general rule, a cold surface—usually glass—is employed because a cold slide won't melt the subject. If it's too cold, however, it may condense other moisture, obscuring the subject. During collection, glass plates coated with oil are preferred, with the oil used to help the crystals stick to the glass. Although there is no accepted "standard method" to photograph snowflakes, Prof. Libbrecht and his colleague Patricia Rasmussen have made huge strides, inventing relatively simple "super-macrophotography" systems, which give a high-quality, repeatable result.

—Tim Kern

GLACIERS COMING & GOING

LEFT: *In spring the frozen sea ice around the Antarctic continent begins to melt and break up. These ice floes, as they are known, move around driven mostly by the wind and tides. Tabular icebergs, which are formed when massive chunks of glaciers and ice shelves break off from the icecap itself, move mostly under the influence of ocean currents. Photo: Jonathan Chester*

ABOVE: *A large mass of ice breaks off the 300-foot-tall ice wall of Alaska's Hubbard Glacier, producing a sound that could be heard for miles. Glaciers typically move very slowly but, on occasion, may move forward as much as several hundred feet per day in what is known as a glacial surge. During a glacial surge in the 1980s, the Hubbard Glacier advanced rapidly across the water, creating an icebound fjord behind it. Marine animals were trapped but the ice wall soon broke, allowing free access once again to the ocean. Photo: Tom Bean*

Humans have been living with glaciers for thousands of years, mostly able to avoid their icy dangers without much worry. In the modern world, glaciers have become major scenic attractions and a welcome thrill for hikers and ice climbers. Scientists have located more than 67,000 glaciers around the world and are using the latest satellite sensing technology to track what is happening to these unique natural features. Increasingly, measurements show that many glaciers are shrinking; yet another potential indicator that the world's climate is warming.

Wherever they are found, glaciers have been created the same way, by the accumulation of snow. As the snow piles up, the weight compresses and freezes it into solid ice when in the presence of frigid temperatures. During warm weather, melting snow on the surface of a glacier channels downward, where colder, internal temperatures refreeze it; some glaciers may remain liquid at the bottom, forming lubrication that helps it slide over surfaces. The tremendous weight of the accumulated snow, ice, and water is gradually influenced by gravity, carrying the whole mass downward, following the slope of the land. At the lowest, outermost edge, warmer conditions melt the ice, cutting into the overall size of the glacier even as more snow and ice is added at the top, upslope surface.

Glaciers that add more ice than they lose are said to be growing; those that lose more than they add are said to be shrinking or receding. If local conditions produce a lot of snow, a glacier may grow even if it is located within a region where warmer temperatures are causing other glaciers to shrink. This is the case in Alaska, where average temperatures have warmed in recent decades—about 7 degrees Fahrenheit over 30 years—reducing the glacier-growing effects typically experienced in this frigid northern region. In some parts of the state, the permafrost is thawing, and most of the state's glaciers are reported to be shrinking, but not all. The Hubbard Glacier, one of the largest in the world, is currently expanding, advancing into the Situk River. The Hubbard Glacier has been creeping forward since the 1890s, and in 1986, a massive ice dam was created on the local fjord, creating a lake about 100 square miles in size before it was breached by the water in the river.

The dramatic Mount Feisland rises to 5,807 feet on Livingston Island, the largest of the South Shetland Islands at the tip of the Antarctic Peninsula. Seen from Baily Head at the mouth of Neptune's Bellows, Deception Island, the Sewing Machine Needles are prominent against the background of sea fog that is common in these icy waters in summer. Photo: Jonathan Chester

In Europe, glaciers in the Alps are estimated to have shrunk by at least 30 percent in area since the mid-1800s, and in the Himalayas, the rate of shrinking has recently accelerated, creating new lakes and causing local flooding. In New Zealand, the shrinkage situation has been noticeable for several decades. In South America, most of the Andes have been similarly affected, and within 15 years, according to some estimates, all of the small glaciers on this continent will have disappeared.

Shrinking glaciers may be a visible sign of a warming climate and future problems for the planet. The volume of water created during melting is one issue, potentially raising the surface of the oceans and causing coastal flooding. In the past five years, the water created by thawing Alaskan glaciers alone is estimated to be twice the amount generated from the ice cap in Greenland, but the amount is still too insignificant overall to impact water levels. Too little water can also produce problems. In some mountainous regions of South America, for example, local drinking water supplies are entirely dependent on glacier runoff. When the glaciers melt away, so too will the water supply.

More immediate than rising water levels is the effect of icebergs created from crumbling glaciers. In 2001, an iceberg the size of Connecticut broke off the Ross Ice Shelf in Antarctica, and growing fissures in the ice shelf suggest another massive fracture may be due soon. Rising temperatures in this area of Antarctica may be one cause; other forces at work include the ocean currents that flow beneath the ice shelf and the fierce winds that blow over it. Effects expected from the next El Niño might also create warmer local conditions in some places along this ice shelf, further accelerating disintegration. But meanwhile, in other regions of Antarctica, it's getting colder, not warmer. Luckily, in this part of the world, even though the icebergs are larger than those created in the Northern Hemisphere, they calve into shallow water and rarely drift out into open waters where they might be a menace to ships.

TYPES OF GLACIERS

TIDEWATER GLACIERS are perhaps the most exciting of all the types of glaciers to watch. These glaciers are mostly found on the coast. Calving ice is their most distinct feature, and separates them from the other glacier types. From the safety of a boat or on shore, watching towers of ice break off into the ocean or a lake is one of nature's most spectacular offerings. Some great examples of tidewater glaciers can be found in Alaska's Prince William Sound and Glacier Bay National Park. These types of glaciers create icebergs that can range in color from deep blue to dark green. The clear icebergs are created over time by the weight of the snow and ice generating compression and forcing out all the air trapped within. Icebergs appear blue due to the fact that ice absorbs all the colors of the spectrum except for blue, which it reflects.

VALLEY GLACIERS are created when alpine glaciers flow downhill and accumulate on a valley floor. The ice flow of these glaciers is similar in character to water flowing in a river. The center flows faster than the edges, and eddies are formed in the bends of this type of glacier. When ice flows at different speeds on a glacier the result is the creation of crevasses, large stress fractures that appear on the top of the glacier. Crevasses allow the glacier to contour or flex over the ground on which it flows.

ROCK GLACIERS are not usually recognized as glaciers, but they deserve to be included in this category because they feature many characteristics found in other types of glaciers. They resemble cirque glaciers in appearance, flow like alpine glaciers, calve like tidewater glaciers, and continue to baffle scientists on their composition. Rock glaciers are composed of a mixture of rock and ice that flows downhill off steep mountainsides.

PIEDMONT GLACIERS are created by at least two glaciers that merge to create a large fan-shaped shelf of ice. Perhaps the best example is the largest glacier in Alaska: the Malaspina Glacier, 850 square miles in total area. Several valley and alpine glaciers from the Bagley Icefield flow into the Malaspina Glacier and fan out at the foot of the St. Elias Mountains. Special features of a piedmont glacier are medial moraines.

airplane, these appear as distinct black lines down the center of the glacier and are created by dirt that the glacier scours from the mountains and deposits in the ice. As the glacier descends like a river, these medial moraines form from the movement created each time it merges with glacier.

GLACIERS can be found high on mountain slopes or plateaus and appear in two forms: cirque ntain. A cirque glacier—this translates from French as "circle" or "ring"—is found in high basins and is often a small circular glacier. Mountain glaciers, on the other hand, flow from plateaus and often feature hanging glaciers or icefalls. As the name implies, a hanging glacier m a steep cliff and often drops glacial ice onto a valley glacier. Icefalls are similar to hanging but instead of dropping ice, they flow like waterfalls. Icefalls are often found on less steep slopes re hanging glaciers are found.

—Rich Reid

GLACIAL ICE MELT GUIDE

Arctic Ocean: Since 1978, the sea ice has shrunk by six percent.

Alaska: The Columbia Glacier is retreating about 115 feet a day as of 1999.

Asia: In the Tien Shan Mountains, the volume of glaciers has shrunk by 22 percent over the past 40 years.

Rocky Mountains: In Glacier National Park, there are less than 50 glaciers today, compared to 150 in 1850.

New Zealand: The Tasman Glacier has retreated about two miles since 1971.

Himalayas: The Dokriani Bamak Glacier retreated 65 feet in 1998.

Europe: Glaciers in the Alps have declined by about 50 percent since 1850.

Argentina: The Upsala Glacier is retreating on average 200 feet a year.

Peru: The Quelccaya Glacier is retreating about 100 feet a year.

—World Watch Institute

BERG HERDING

Icebergs that drift around in the North Atlantic pose serious risks to shipping and offshore oil rigs. But modern technology helps keep track of the drifting floes—satellite images and sonar from search planes help locate and follow their movement. Smaller bergs can be redirected with blasts from water cannons; powerful boats with towlines are used to divert bigger floes into safer routes.

GUIDE TO ICEBERG SIZES

Term	Height	Length
Growler	less than 3'	less than 16'
Bergy Bit	3–13'	16–46'
Small	14–50'	47–200'
Medium	51–150'	201–400'
Large	151–240'	401–670'
Very Large	240'+	670'+

LEFT: *Small icebergs in the Southern Ocean are often tossed and turned by oceanic waves until they are eroded into fantastic shapes. This melting iceberg was photographed in Crystal Sound, just south of the Antarctic Circle, near the Antarctic Peninsula. Photo: Joan Myers*

RIGHT: *Ice caves such as this frequently form in the Erebus Ice Tongue, a floating glacier that spills off the side of Mount Erebus on Ross Island, Antarctica, and stretches into the waters of McMurdo Sound. When a crevasse forms, sea water pours in and freezes, forming the floor of an ice cave. Some of these caves can be quite large, reaching back into the glacier for more than 100 feet. Entering one is like walking into a silent, crystal cathedral with the walls and ceiling composed of hard, brittle ice in every shape and shade of blue. This cave had a ceiling 24 feet high. Photo: Jim Mastro*

LEFT: *Spawned by a glacier in the Antarctic, this iceberg drifted north of the South Sandwich Islands and was quickly being destroyed by the inhospitable climate of the Southern Ocean. Waves crash against the iceberg and cascade 20 feet into the cold waters below. The blue color is the result of dense glacial ice refracting light so that all colors except for blue are absorbed. Pintado petrel birds float below looking for tidbits of food stirred up by the turbulent seas. Photo: Colin McNulty/Small World Images*

RIGHT: *Researchers take an October dip into the Ross Sea to explore this iceberg, run aground at a depth of 84 feet, just south of Cape Evans on Ross Island. In the Antarctic, all life revolves around the ice. The iceberg shelf is covered in ice algae and loaded with bright red amphipods grazing for food, who in turn attract large numbers of hungry ice fish. Larval fish by the thousands rise and fall along the anchor ice. Photo: Norbert Wu*

FOLLOWING PAGES: *Altocumulus clouds are reflected in the iceberg-strewn ocean at the mouth of the Weddell Sea. Close to the Antarctic coast, it can be very calm, especially inside the belt of pack ice that can dampen the swells. Photo: Jonathan Chester*

CLIMATE CHANGE

GLOBAL WARMING, GLACIAL MELTING, DESERTIFICATION, AND POLLUTION

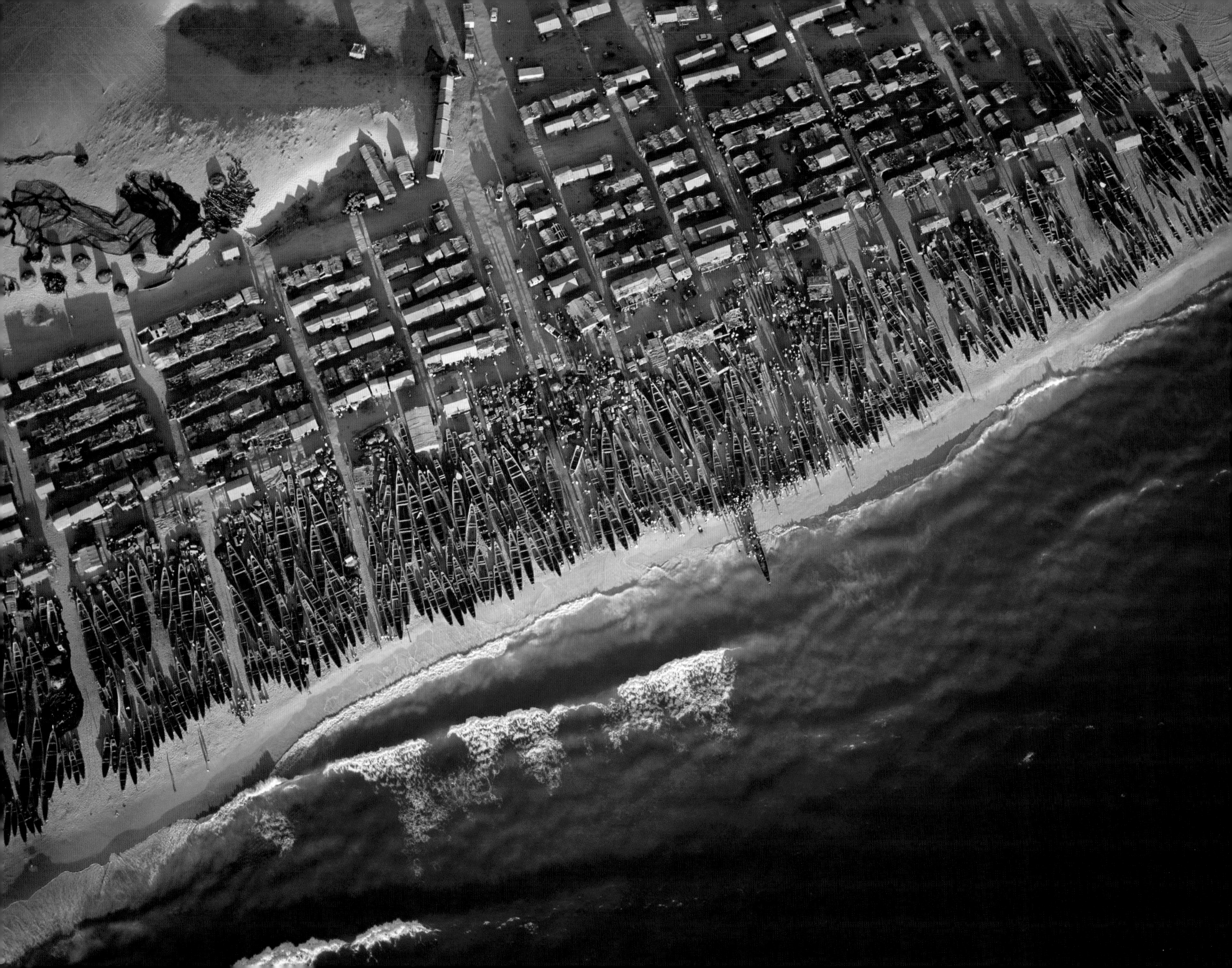

The old joke in meteorology school use to be: "Climate is what you expect; weather is what you get." Trouble is, climate, once thought of as an unchanging mean about which fluctuations would bounce, is now anything but unchangeable, even on the scale of humankind's existence on earth, and now even for an individual lifetime. Truth be said, climate has always been changing, long before humans showed up and picked fights with wooly mammoths and saber-toothed tigers. Continents have wandered—witness oil under the Arctic sea and coal beds in the Antarctic. Meteorologists once attempted to measure the "solar constant" of incoming energy which actually fluctuates sufficiently to alter climate. Volcanic eruption dust clouds cause "years without a summer." Once humans abandoned the earth-centric notion of the universe, we were comforted by a predictable elliptical orbit around the sun, except that the ellipse varies on the scale of tens to hundreds of thousands of years while the planet wobbles on its axis like a crazy top, inducing all manner of climate cycles.

When Homo sapiens showed up and started burning fossil fuels and changing the landscape, these were seemingly small change in the grand scheme of things. Then late-19th-century rises in carbon dioxide levels caused a few scientists to start thinking about greenhouses. But how could mere humanity change the climate of an entire planet? Surely Mother Earth would be more resilient than that? We once thought we couldn't pollute a river—until Ohio's Cuyahoga caught fire. We once thought we couldn't pollute our atmosphere—until acid rain soured our lakes, giant brown clouds of Asian dust crossed the Pacific, and smog blobs choked the air over the eastern U.S. on hot summer days. We couldn't pollute the oceans—until hypoxic "dead zones" appeared at the Mississippi's mouth, and six-pack holders floated in mid-Pacific along with thousands of rubber duckies from container spills off giant cargo vessels.

But change climate? The upward march of atmospheric carbon dioxide levels is unquestioned and unrelenting as we burn ever more fuel. Beyond CO_2, heat-trapping methane emerges from pastures of grazing cows, bucolic rice paddies, and melting arctic permafrost. Mother Earth—or the Sun-Earth-Ocean System as some call it—is responding to both natural and human-induced changes in ways both measurable and still unpredicted. Humanity is clearly conducting an uncontrolled experiment on its own atmosphere. Sea level creeps upwards, Greenland meltwater flows ever faster, spring comes earlier and earlier in mid-latitudes, and polar bears searching for ice floes become iconic images, just as did the breathtaking portrait taken from the Apollo capsule of Spaceship Earth rising over the moon. Our spaceship. The one on which we must all keep traveling, at least for a long, long time.

The stern prose of the Information Statement on Climate Change by the American Meteorological Society tells it all: "Despite the uncertainties noted above, there is adequate evidence from observations and interpretations of climate simulations to conclude that the atmosphere, ocean and land surface are warming; that humans have significantly contributed to this change and that further climate change will continue to have important impacts on human societies, on economies, ecosystems and wildlife through the 21st century and beyond . . . [I]n the next 30 years, increasing air temperatures will reduce snow pack, shift snow melting, reduce crop production and rangeland fertility and cause continued melting of the ice caps and sea level rise." What have we children of earth done? Is this any way to treat our Mother?

—Walter Lyons, CCM

Fishing boats line the once deserted shore of what is probably the fastest growing city in Africa. Before independence in 1960, Nouakchott was a small coastal oasis supporting 15 families in mud houses. Now, over 700,000 people have flocked to this capital city to seek wealth and refuge from desertification of their traditional grazing lands. As their herds of livestock vanished, fishing has become a major part of the national economy. Mauritanian waters were once the richest fishing grounds in Africa, but with rampant overfishing by large offshore trawlers, fish populations are now only 30 percent of what they were 20 years ago. Photo: George Steinmetz/Corbis

THE GREENHOUSE EFFECT

Radiation that strikes the earth from the sun is transformed into heat. Heat radiates back into space through the atmosphere. If too much heat escapes into space, the earth can get too cold; if too much is trapped and prevented from escaping, the earth can get too hot. The regulating factor is gases in the atmosphere—molecules that allow incoming radiation to pass unimpeded but absorb the heat radiation bouncing back. These naturally occurring compounds are called greenhouse gases and include water vapor, carbon dioxide, and methane among others.

Human activity alters the quantity of these gases, with carbon dioxide, for example, produced by burning fossil fuels. Man-made gases, such as chlorofluorocarbons, also alter the natural mix. The concentration of carbon dioxide and methane are both at the highest levels in hundreds of thousands of years, due in part to an increase in human activity. Yet even as human activity generates more CO_2, increasing the greenhouse effect and raising the temperature of the earth, the increased temperature and quantity of CO_2 will speed up the growth of vegetation, increasing the amount of greenery that absorbs CO_2.

On the other hand, warming will melt huge areas of permafrost in northern latitudes, releasing large amounts of methane into the atmosphere, adding to the greenhouse effect. Compounding the issue is the problem of clouds. As the temperature rises and moisture levels in the air increase, more clouds will form. But clouds not only trap surface heat, raising temperatures, they radiate solar radiation, lowering the amount of heat available to the surface. No one yet knows what the net effect will be.

—*Kim Long*

LEFT: *A dried-up river surrounds a village during a drought in the Sahara Desert, Mali. Photo: Charles & Josette Lenars/Corbis*

RIGHT: *Tuvalu, a Polynesian island nation, is at the front line against global warming. Fifteen feet above sea level at the highest point, the rising sea is putting the population of 10,000 at risk. It is likely that Tuvalu will be the first country to disappear as a result of climate change. Already during the highest tides, sea water is forced up through the porous coral atoll, flooding many low-lying areas. Photo: Ashley Cooper/Corbis*

GLOBAL WARMING:
THE GOOD, THE BAD, AND THE UGLY

LEFT: *The severe pollution in Beijing, seen here at sunrise, illustrates China's surmounting environmental challenges. According to a report from the World Bank, China has 16 of the world's 20 most polluted cities, and its carbon dioxide emissions have nearly doubled since 1985. Photo: Dean Conger/Getty Images*

ABOVE: *Weather extremes from flooding to drought are clear indications that the average global temperature has risen about one degree Fahrenheit over the past 100 years. Photos: Bob Firth/Firth Photobank (top left); Ed Darack (top right); Fred Hirschmann (bottom left); and Rich Reid (bottom right)*

In the past 100 years, the average surface temperature of the planet has increased by about one degree Fahrenheit. The climate of earth is always changing, usually slowly over time, but this temperature change has attracted widespread concern not only because it may be partly caused by human activities, but also because it has widespread implications for life forms, including humans. If it continues increasing at its present rate, within 100 years the average temperature could rise by another two to four degrees Fahrenheit. This variation would be unnoticed if it were a daily fluctuation, well within the range of local weather conditions, but as an average increase representing the whole planet, each rising degree has greater impact. The difference of only a few degrees in average temperature in an ecosystem, for example, can turn a prairie into a forest. Depending on the scenario, the effects could vary from insignificant to catastrophic.

While average temperatures change, there is a different change expected for extreme temperatures—the highest highs and the lowest lows. At least one study concludes that global warming will mean more extreme high temperatures but less extreme low temperatures. The average warming also masks the reality of what may change in a 24-hour period. Rather than the overall temperature rising, the nighttime low temperatures would increase more than the daytime highs, that is, nights would be warmer more than the days would be warmer. And during the day, daytime highs would not rise as much as daytime lows, driving up the averages but not making things more uncomfortable.

As the oceans rise along with the temperature, coastal residents may suffer from too much saltwater. But the opposite effect could plague large areas: too little freshwater. Many cities, particularly in the American West, rely on fresh water runoff from mountain snowpacks, supplies that would be reduced if less snow falls. This might only be a seasonal condition, however, as regular rains during the traditional rainy seasons are also expected to increase. For the Pacific Northwest, some experts predict more flooding in wet seasons and more forest fires in dry seasons.

GREENING TRENDS

While added warmth may result in flooded urban areas along the coast—a potentially catastrophic economic condition—longer growing seasons farther north in the United States and Canada would improve the productivity of farms. In some cases, given enough rain, farmers located as far north as the upper Midwest could be able to grow two crops a year instead of one, a beneficial effect of warming. This is because the length of the cold-weather season—when freezing temperatures inhibit plant growth—would shrink, allowing earlier plantings. Since the 1950s in the northeastern states, for example, the beginning of the frost-free growing season has already shifted forward by 11 days. For the United States as a whole since 1910, there are now slightly fewer days that are below freezing.

It's not just the temperature change that would benefit agriculture, however, but also the carbon dioxide increase in the atmosphere. Plants utilize carbon dioxide in their growth cycle, where it acts like a fertilizer, increasing the rate of growth, the ultimate size of the plant, its ability to withstand severe conditions, and the amount and size of seeds and fruits produced. With CO_2 levels already rising, some reports have confirmed that the production of major food crops—including grains, fruit, vegetables, beans, other legumes, and root crops has already increased, perhaps 10 percent more than would be expected with improvements in farming ASA (American Soybean Association) technology. And since plants consume CO_2, the more plants, the more CO_2 consumption, theoretically moderating the greenhouse effect by removing CO_2 from the atmosphere. Unfortunately, it's not just valued agriculture crops that will thrive in the CO_2-rich environment. Weeds will, too.

While many people might feel more comfortable if their local climates were warmer, there are negative effects for human health as the temperature rises. Just as the mix of plant species would adapt to take advantage of the added warmth, so too would many animal pests. Agricultural enemies—such as aphids, boll weevils, and termites, for example, would thrive in destructive numbers farther north than the zones in which they are now confined. The decrease in freezing temperatures would allow more mosquitoes to breed over longer periods of time, and malaria would be a problem far beyond its present boundaries, where it is naturally limited by cold temperatures. One estimate suggests that by the year 2100, the proportion of the world's population exposed to malaria would rise to 60 percent from its present level of only 45 percent. And that's only one disease that is linked to mosquitoes and climate; West Nile virus, which first popped up in New York City in 1999, can cause encephalitis in humans and could also thrive in a warmer North America. Another health menace comes with higher-than-normal precipitation and is already a problem with El Niño-influenced weather patterns in the Southwest: this weather triggers a population explosion of deer mice. When these mice experience a boom, the hantavirus that they carry is more likely to be transmitted to humans, leading to local and regional epidemics.

Scientists are divided on the issue of severe weather and global warming. It's mostly a problem of lack of information, with too little known about the complex interactions between slow climatic changes and local weather conditions. Only time will tell. We can only hope that if severe weather changes are indeed the result of man's actions, we will not respond too late.

—*Kim Long*

A man wearing garbage bags on his legs struggles through the floodwaters on the sidewalk along the Canale Della Giudecca in Venice, Italy. Strong winds and a rising sea level often flood the city. Photo: Michael Hanschke/dpa/Corbis

RISING SEA LEVELS

Due to global warming, temperatures at the poles will experience the greatest change, perhaps as much as six to eight degrees Fahrenheit. The west Antarctic ice sheet, unlike the ice on the eastern side, extends out into the ocean. A warming trend would not only turn more of this frozen mass into water, raising ocean levels, but also warmer conditions would contribute to fractures and slides, dumping huge chunks of ice into the water, creating destructive waves and a rise in water levels throughout the globe.

Evidence found shows that this has already happened in the past, as long as a million years ago. If the entire western Antarctic ice sheet melted, which is not likely, it would produce enough new water in the world's oceans to create an estimated rise of 20 to 33 feet, not counting the lesser effect of melting glaciers.

Along the East and Gulf coasts of the United States, this would flood vast areas of now-occupied coastline, including most of the Mississippi Delta and the lower one-third of the state of Florida. A more reasonable expectation over the next 100 years is for a sea-level rise of only one to three feet, mostly due to thermal expansion—the added volume created when water is heated—which is still enough to cause major problems for coastal residents.

—Kim Long

DROUGHT VS. FLOOD

One of the most talked about characteristics of global warming is a probable rise in the level of the ocean, causing serious problems for cities and residents of many coastal areas. But even more problems related to water can be expected. With the rise in temperature comes an increase in rainfall in some areas with the ultimate outcome being severe flooding.

The current trend for rain in the U.S. is for more days with two inches or more of precipitation, particularly in the southwestern, midwestern, and Great Lakes states. And more heavy rain may lead to more flooding. And across the country, even if there isn't more rain and more flooding, humidity is increasing.

As for drought effects, most are expected to be regional, not global, and while some areas may suffer for lack of rain, others will get more than normal. One climate model based on global warming also shows that in the Northern Hemisphere, soil moisture levels in the mid-latitudes would decrease, but only during the summer months; in the winter, the soil would have more moisture. Some of the climate models applying global warming trends to the U.S. project droughts that are more frequent, longer in duration, and more intense. In any case, even as the average temperature of the earth has crept up over the past 100 years, there has yet been no change in the traditional pattern of drought conditions in the United States.

—Kim Long

TOP RIGHT: *Hundreds of houseboats are lined up as shikaras (boats) carrying tourists pass in front of them on Dal Lake. The lake, which once covered 12 square miles, has shrunk to half that size over the past four decades because of accelerated global warming, and silt and weed development. Photo: Altaf Qadri/Corbis*

BOTTOM RIGHT: *An elderly Tuvaluan man watches helplessly as waves batter Funafuti Atoll during one of the highest tides of the year. Photo: Ashley Cooper/Corbis*

OPPPOSITE PAGE: *While earth's climate has undergone cooling and warming cycles in the past, the rate and magnitude of change today has not occurred since human civilization began. If the current speed of warming persists, scientists predict the glaciers in Glacier National Park will be completely gone by the year 2030. The number has already been reduced from 50 in 1968 to 27 today, many of which are mere remnants. Photo: James Randklev/Corbis*

WHAT'S HAPPENING TO THE ICE?

LEFT: *Grounded icebergs and bergy bits are carved by wave action and melt into mushroom-shaped objects with the rise and fall of the tides. Spray and meltwater refreeze in dramatic icicles when the temperature drops each evening. Photo: Jonathan Chester*

ABOVE: *Atolls of the Maldives. Photo: Michael Fiala/Corbis*

For most people, the thought of glaciers and polar ice caps brings on visions of vast, frozen wastelands with polar bears or penguins. It's a great topic for a PBS special, but the great ice fields have little importance in most people's minds. For others, however, it can be a life or death matter. Consider the ill-fated *Titanic*, whose doom was sealed by a Greenland iceberg in the North Atlantic. Or for a more current concern, consider the small Indian Ocean nation under threat from Arctic ice, the Republic of Maldives.

The Maldives is a chain of more than 1,000 coral islands stretching perhaps 750 miles off the southwest coast of India. About a quarter million people inhabit only 200 of the islands, none of which is more than five square miles in area. Most are off limits to outsiders, although several dozen mostly tiny islands are set aside for tourism. With beautiful palms, wide sandy beaches, stunningly blue waters, addictive scuba diving, and a myriad of colorful fish, the Maldives would seem the most carefree place in the world. But the Maldivian government has plenty to worry about—the world's glaciers and polar ice fields!

The average elevation in this tropical paradise is less than two meters, about 6½ feet. A relatively small rise in the average sea level, six inches, could have disastrous results for the islands, significantly reducing beachfront acreage and tourism dollars. A large rise of three feet, and it would pretty much be "bye-bye Maldives."

But do they really have anything to worry about? You bet. According to some recent research, not only the Maldives but also parts of the Netherlands, the historic city of Venice, and many other low-lying coastal areas around the world could be partially inundated in the next century. The International Panel on Climate Change has predicted that by the year 2100, the average sea level would rise by a foot and a half over current levels. While only part of that would be directly attributable to melting ice sheets and glaciers, more than half would be due to the thermal expansion of warming waters, an indirect consequence of glacial melting.

In Greenland, the glaciers are thinning, according to a recent NASA study. "The excess volume of ice transported by these glaciers has had a negligible effect on global sea level thus far, but if it accelerates or becomes more widespread, it would begin to have a detectable impact on sea level," said principal investigator Bill Krabill. "Why they [glaciers] are behaving like this is a mystery," said Krabill, "but it might indicate that the coastal margins of ice sheets are capable of responding quite rapidly to external changes, such as a potential warming of the climate."

It isn't just the Greenland glaciers that are losing ice. Two recent University of Colorado (CU) studies point to changes in polar ice and glaciers at lower latitudes. Researchers at CU's National Snow and Ice Data Center (NSIDC) and the British Antarctic Survey have reported that portions of the Antarctic ice shelf are breaking up due to decades of higher temperatures. Radar satellite imagery has confirmed that two Antarctic ice shelves, called Larsen Band Wilkins, are in "full retreat," having lost nearly 3,000 square kilometers of their area in the year from spring 1998 to spring 1999.

These Antarctic ice shelves are likely hundreds of years old, and it is a completely natural process for the ice to break up. Much larger ice sheets have broken off the Antarctic ice shelf. From 1967 to 1978, a huge sheet of more than 7,000 square kilometers was tracked by satellite before it broke up and melted away. And in 1956, officers of the USS *Glacier*, a Navy icebreaker, reported a gargantuan ice sheet, 60 miles wide by 208 miles long, about 150 miles west of Scott Island in the Ross Sea. This behemoth was twice the size of Connecticut. Still, the recent rapid retreat of the Larsen Band Wilkins shelves is remarkable.

"We have evidence that the shelves in this area have been in retreat for 50 years, but those losses amounted to only about 7,000 square kilometers," said David Vaughan, a researcher with the Ice and Climate Division of the British Antarctic Survey. "To have retreats totaling 3,000 square kilometers in a single year is clearly an escalation. Within a few years, much of the Wilkins ice shelf will likely be gone."

In a separate study, also from the University of Colorado and released in May 1999, researchers found that glaciers overall are in retreat, presumably due to warming global temperatures. "In the last century, there has been a significant decrease in the area and volume of glaciers, especially at mid- and low-latitudes," said Professor Emeritus Mark Meier of the geological sciences department at the University of Colorado.

Meier cited major retreats of glacial ice on Africa's Mount Kilimanjaro and Mount Kenya and in Europe, where glacial ice loss has been roughly 50 percent in the past century. He also reported on Montana's Glacier National Park, where the namesake ice formations may be gone altogether within 100 years.

While glaciers outside Greenland and Antarctica amount to only about 6 percent of the world's total ice mass, they contribute more to sea level rise than do the polar ice sheets, Meier noted. "Glacier changes show strong regional differences," he said. "While the Arctic ice caps and glaciers show little change, there is strong wastage of mid-latitude glaciers, and small continental glaciers are disappearing." He also pointed out that this is causing significant increases in the flows of some rivers.

Only time will tell the full extent of ice loss and its effect on sea level rise, but there can be little doubt that it is related to a worldwide rise in temperatures. "During the past several decades, ice wastage and global sea rise are moving pretty much in step," Meier concluded. Meanwhile, the Maldives and other susceptible coastal areas can only formulate contingency plans, and pray for a return of the ice.

—Larry Sessions with Jim Scott of the University of Colorado, who contributed significantly to this article

RIGHT: *These two images display the extent of Arctic ice melt over a nine-year period. The bottom image was taken January 1, 1990, and the top on January 1, 1999. Photos: NASA*

FAR RIGHT, TOP: *Small icebergs and floes from the end of a glacier float in a meltwater lake at Glacier National Park in Montana. Photo: Kennan Ward/Corbis*

FAR RIGHT, BOTTOM: *The glacier at the top of Mt. Kilimanjaro, Africa's highest peak, is receding—another visible casualty of global warming. Photo: Kazuyoshi Nomachi/Corbis*

MEASURING ICE

Scientists use a variety of methods to measure the Arctic icepack. Submarines roving in the sea under the floating ice direct their sonar upward to measure ice thickness; the reflected signal changes according to the thickness. Ice breakers also participate by measuring the thickness of pieces of ice broken and scattered by their passage. Records from previous voyages are compared to show changes due to weather and over time. The best measures for ice coverage in the Arctic sea come from satellites, which can distinguish the long-wave radiation emitted by sea ice from heat that is emitted by water.

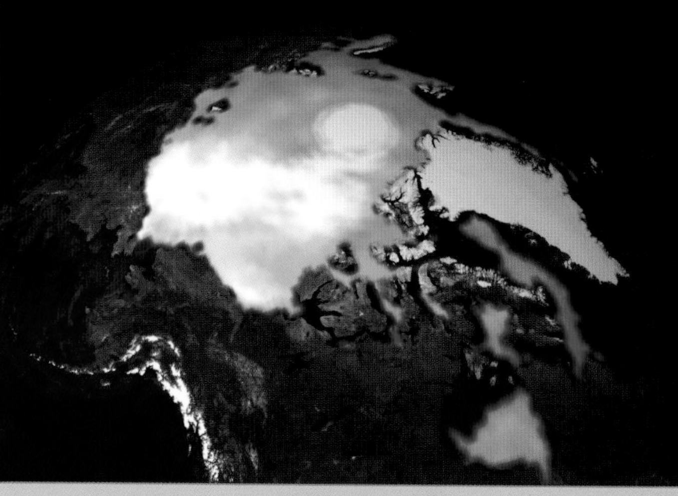

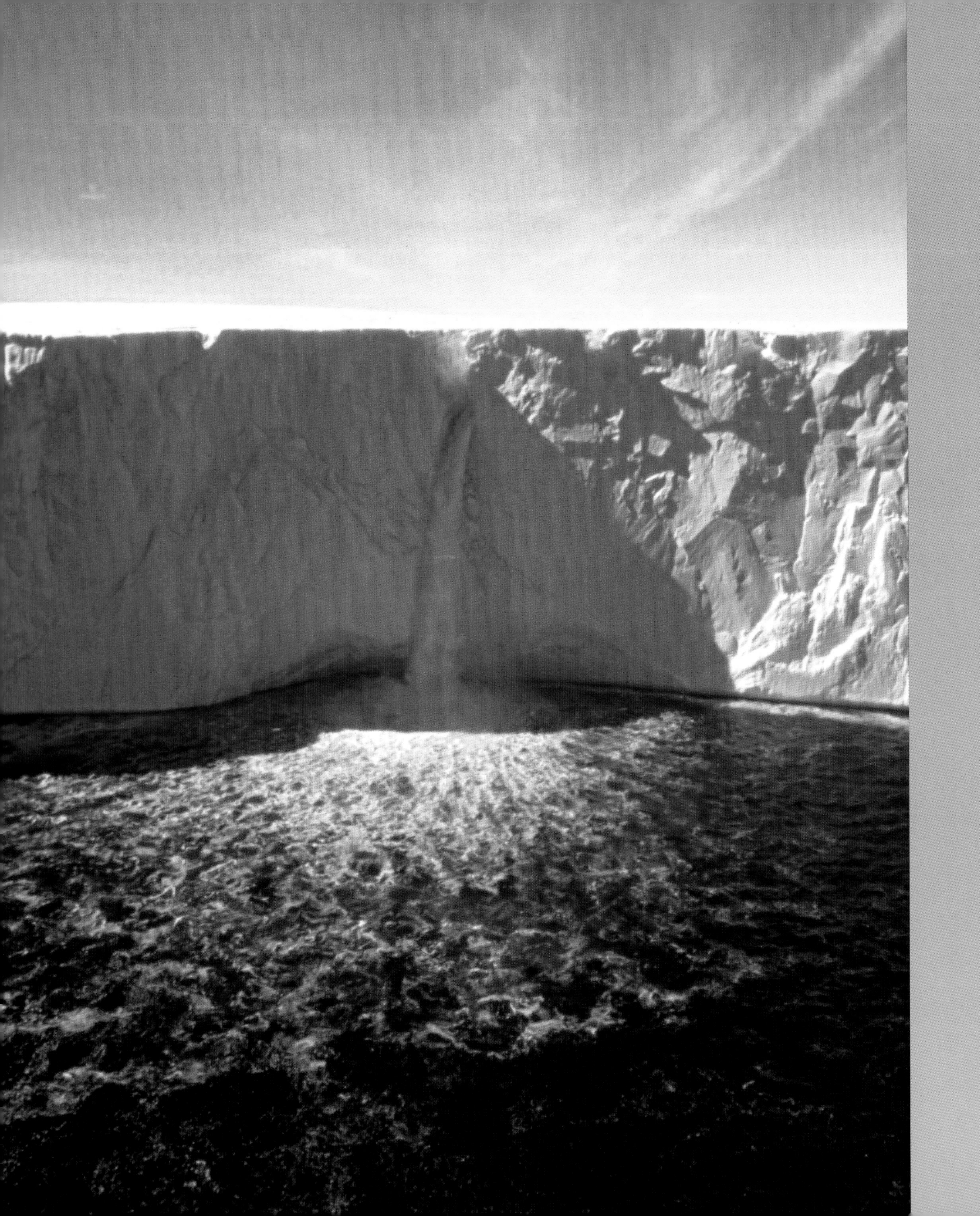

ON THIN ICE

"Strange as it seems, the Arctic ice is melting at so precipitous a rate that scientists now believe that the Northwest Passage may be navigable by regular ships for part of the year, or even all of it, in as little as 10 to 15 years."—*Toronto Globe and Mail, February 5, 2000*

The explorers who exhausted themselves searching for the Northwest Passage might be heartened to know they were not wrong, just 500 years too early. Perhaps no area on earth will be as affected by global warming as much as the Arctic. The Intergovernmental Panel for Climate Change says temperatures have increased nine degrees Fahrenheit on average in the Arctic over the past century, and they predict temperatures will continue to rise more there than anywhere else on earth.

According to the Nansen Environmental and Remote Sensing Area in Norway, the ice cover decreased about three percent every 10 years between 1978 and 1996. The thickness of the ice decreased an average of 42 percent in all the main areas of the Arctic Ocean between the periods 1958–1976 and 1993–1997.

The perennial ice—ice that has survived one summer or more—has shrunk more than twice as fast as the total ice cover, according to Nansen scientists. Put another way, as the winter freeze adds a little less ice each season, the net effect of seasonal melting is that the thicker, semi-permanent base gradually decreases.

Two University of Bergen researchers summed it up this way: "What is completely certain is that if the trend in the ice cover and the ice thickness does not level off soon, the perennial ice in the Arctic will disappear by the end of this century. In other words, the entire Arctic will be free of ice during the summer months."

— *Randy Welch*

LEFT: *There is sufficient melting on the surface of the Sorsdal Glacier in summer to have streams coalesce into a river ending in a spectacular waterfall that plunges hundreds of feet into the ocean. The Sorsdal is one of the key outlet glaciers that contribute to the drainage of the East Antarctic ice sheet. Photo: Jonathan Chester*

RIGHT: *Closed pack ice occurs in the spring when wind and currents break up the fast ice and move the frozen plates into a tight band of floating ice. These might be anywhere from a few hundred yards to miles in extent. This tight formation makes passage impossible for all but an icebreaker, but ice charts that are compiled daily from satellite imagery give ships' captains the means to avoid or minimize the problem. Photo: Jonathan Chester*

DESERTIFICATION

LEFT: *A sandstorm on a dirt road between Douentza and Timbuktu in Mali. Photo: Remi Benali/Corbis*

ABOVE: *A Chinese villager waters a tree planted to combat desertification on a hill. China, the second largest source of greenhouse gases after the United States, is one of the world's worst affected countries by desertification. Photo: Diego Azubel/epa/Corbis*

Study a physical map of the world. As you follow the tropics of Cancer and Capricorn, 23.5 degrees on either side of the equator, you will see a brown band of deserts circling the planet. They lie in the "horse latitudes," where constant high-pressure systems separate the westerly and trade winds, driving away the rain clouds. Some of those dry lands, like the Atacama of Chile, the Namib and Kalahari deserts of southern Africa, and the western Australian desert, are the result of cold ocean currents that divert rain-laden air away from coastlines. Others, like the Mojave and Sonoran deserts of California, Arizona, and Mexico and the deserts of central and eastern Australia, are caused by the "rain-shadow effect," through which coastal mountains milk rain from the air before it passes inland. Still others, like the Gobi and Taklamakan deserts of Mongolia and China, are simply so far away from the ocean that the winds lose any moisture they may hold long before reaching the distant continental interior.

The world's desert systems are harsh environments. By definition, a desert receives less than 10 inches of unevenly distributed rain throughout the year, though it need not suffer extreme hear. (Antarctica, for example, is a desert where rain never falls and no vegetation grows.) These deserts now cover some 20 percent of the planet's surface, a figure that grows each year, thanks to the phenomenon of desertification, a process of soil erosion and land degradation that occurs when land that normally receives little rain is stripped of whatever vegetation it has.

Desertification results from several natural processes that are harmless enough one by one but can play ecological havoc in combination. One is the ongoing process of global warming, which is altering weather patterns around the world, bringing drought to once temperate zones. Another is the desiccating El Niño weather system in the Pacific, which has been prevalent for much of the past decade and has led to a virtual stop of summer rain in the deserts of North and South America.

But the most powerful agent of desertification is humankind. With the growing world population, formerly marginal areas on the fringes of deserts are becoming more heavily settled. With humans come livestock, which devour the already scant ground vegetation; taller trees and shrubs are cleared away for fuel wood. The removal of plant life means that when rain falls, it cannot penetrate the dry soil, once broken by plant roots; instead, it runs off the surface toward low ground in a process that hydrologists call "sheet flooding." In the last quarter century, according to United Nations statistics, at least 128,000 people have died as a result of such floods.

Desertification has caused tremendous social change in affected areas, especially in sub-Saharan Africa and Central America, where famine is now a constant danger and massive flooding and mudslides follow even modest rainfall. In these areas, desertification has caused the destruction of rural agriculture and a massive migration of country people into already crowded cities.

Desertification has emerged as a major environmental problem in some unlikely areas. In the deserts of the United States, areas that have been intensively grazed and farmed have grown a thick skin of salts and other minerals, making the land useless for further agriculture. In Beijing, dust storms from the nearby Gobi are a regular hazard, while the Gobi's sand dunes advance toward the city at a rate of as much as 15 miles a year. In Italy, Spain, and elsewhere in the Mediterranean, intensive olive farming has led to soil erosion and canyon-cutting, and great areas of land are now unsuitable for agriculture.

Desertification is not unstoppable, but containing its spread will require massive international efforts and cost trillions of dollars. Any measures to halt its growth will involve continued economic hardship for the people most affected by desertification, for they include putting an end to livestock grazing and irrigated agriculture until plant cover has returned to a denuded stretch of ground, a process that can take decades. International aid organizations are working to convert farmers and herders in places like the Sahel of West Africa and northwestern India into modern-day Johnny Appleseeds, planting hedgerows and windbreaks to halt the advancing sands. The fate of hundreds of millions of lives and of millions of acres hinges on their success or failure.

—Gregory McNamee

ABOVE: *In Ningxia province, the lack of water has forced peasants to collect snow to stock in underground reservoirs. The central government launched a program called "Mother Cellar" which aims to build underground storehouses capable of holding water for several months. The desertification of this region is extreme and there are no wells or trees. Snowfall is greeted with relief by the locals. The only form of revenue for families is tree planting with saplings ordered by the government for reforestation purposes. Other than the water from the underground storehouses, the only way to access water is for peasants to buy water delivered by trucks. Photo: Yves Gellie/Corbis*

THE BURIAL OF BEIJING

For most of the people of the world, desertification is something that happens far away from home. But in Beijing, the phenomenon is much more real, creating serious problems in this massive city, home to about 11 million people. Sand, dirt, and dust originating in the shifting deserts of northwestern China are now frequent visitors here. Many residents now dread the spring, when annual conditions trigger multiple sandstorms. In 2000, as many as a dozen sandstorms struck the city during the spring months, and for some observers, the intensity seemed worse than in previous years. During such storms, visibility is reduced and breathing can be difficult; during peak storm periods, airports are closed and hospitals fill with casualties of the wind-borne muck.

Not only is this "muck" a visible sign of desertification, the sand dunes themselves are also on the march. Scientists measure their progress southward toward Beijing at about 2.1 miles per year—some measurements put the movement as high as 15.5 miles a year. The small city of Huailai, which is currently being enveloped by these dunes, has recently had a name change. It is now called Sand City.

The Chinese are not just studying this menace; they are making plans to protect Beijing. One effort is the planting of large numbers of trees about 112 miles north of the city, in Hebei province, where desertification is particularly intense. Bans are also in place to prevent the harvesting of black moss, a natural ground cover, and to limit the cultivation of ground for agriculture in sensitive areas.

—*Gregory McNamee*

LEFT: *A worker carries hay used to prevent desertification at the edge of the expanding Maowusu Desert. Sand dunes are encroaching as a result of grazing livestock and constant drought in one of China's poorest regions. China spends about 2 billion yuan ($280 million) annually to prevent desertification, which affects 2.64 million square kilometers. Twenty percent of China's total landmass is desert. Photo: Michael Reynolds/epa/Corbis*

THE WORLD'S LARGEST DESERTS

Sahara	Northern Africa	3.5 million sq. mi.
Gobi	China	500,000 sq. mi.
Patagonia	Southern Argentina	300,000 sq. mi.
Rub al-Khali	Southern Arabia	250,000 sq. mi.
Kalahari	Southern Africa	225,000 sq. mi.
Chihuahuan	Mexico/southwestern US	175,000 sq. mi.
Great Sandy	Western Australia	150,000 sq. mi.
Great Victoria	Southwestern Australia	150,000 sq. mi.
Gibson	Western Australia	120,000 sq. mi.
Kara Kum	Turkmenistan	120,000 sq. mi.
Syrian	Northern Arabia	100,000 sq. mi.
Thar	India	100,000 sq. mi.

Desert areas are very approximate because clear physical boundaries may not occur.

TOP LEFT: *Signs of desertification are apparent in fields located on the outskirts of the city of Segou, Mali. Photo: Remi Benali/Corbis*

BOTTOM LEFT: *A dead bull on a dry plain in central Mali is evidence that life in Africa's Sahel is becoming increasingly difficult. Desertification and rising temperatures due to climate change are making farming nearly impossible and the survival of livestock of critical concern. Photo: Nic Bothma/epa/Corbis*

RIGHT: *A Touareg man fills jerrycans with water from one of the few wells in working order. Thirty years of drought have seriously affected the equilibrium and daily lives of northern tribes in Mali. The progressive disappearance of watering holes and the advance of the desert have pushed tribal populations to installations on the periphery of towns. Photo: Bruno Fert/Corbis*

CHINA, COAL & GLOBAL POLLUTION

LEFT: *A coal power plant spews toxic fumes into the atmosphere in Hebei province, China. Photo: Liu Liqun/Corbis*

ABOVE: *A migrant worker shovels coal at a depository in the outskirts of Beijing. By the end of 2007, Beijing will completely ban all such small privately operated coal depositories in efforts to reduce the capital's notorious air pollution before hosting the 2008 Olympics. Despite government pledges to curb environmental pollution, China's economy heavily relies on coal to meet growing energy demands. Photo: Michael Reynolds/epa/Corbis*

What goes up, the adage has it, must come down. It's an old law of the universe, and universally recognized. Even so, when in 2004 a professor of atmospheric sciences at Harvard University, Daniel Jacob, published the results of a study of a local weather phenomenon, it surprised many New Englanders. Jacob announced that a plume of polluted air that had been hanging in the region's skies had drifted all the way from China, barely affected by the moderating and buffeting winds of the jet stream. The smoky cloud's endurance was bad enough, but it was also full of mercury and other highly toxic materials given off by the burning of coal, among them sulfur dioxide, nitrogen oxide, and sooty particulates.

Two years later, in April 2006, a thick cloud of smoke that had been hanging over heavily industrial northern China blew away, as if unmoored from earth. It settled over Korea for a few days, enveloping Seoul in a thick blanket, and then sailed across the Pacific, making landfall in California a few days later. The cloud was so thick that satellites could track it from far above earth, as if it were a landmass. Only when it hit the tall barrier of the Sierra Nevada did the cloud begin to break up, shedding sulfur, mercury, and other by-products of coal burning into Lake Tahoe.

Coal smoke has been an atmospheric problem for as long as coal has been used as a fuel. Londoners of a certain age still remember the deadly winter inversions that once choked the city in coal smoke and fog, giving rise to the word "smog." The "forest death" that visited so many Old World woodlands in the 1970s and 1980s was largely the product of the coal-burning industries of northern Europe, just as the "acid rain" of the Eastern Seaboard owed its existence to coal-burning plants there. Recognizing the dangers wrought by coal, the nations of the industrialized world developed filtering systems that scrubbed the smoke emerging from industrial smokestacks, even as most consumers switched over to natural gas and other cleaner heating fuels.

Yet today, a third of America's lakes are so polluted with mercury that warnings are issued against eating fish taken from them. One-half of that mercury, it is estimated, comes from China, whose factories and power plants release nearly 600 tons of it into the atmosphere every year, along with 22.5 million tons of sulfur.

RESPIRATORY ILLNESS DRASTICALLY ON THE RISE

By 2010, as many as 600,000 Chinese will die of respiratory illnesses each year, a marked increase over today's rates, which already amount to nearly 400,000 deaths. Asthma, a chronic inflammation of the airway once fairly rare in China, now afflicts millions of people there. Rates of asthma also have risen elsewhere in the industrialized world, and it accounts for tens of thousands of deaths annually, especially on hot days, which exacerbate the condition.

Some evidence suggests that changing climatic conditions are conducive to new epidemic diseases such as SARS and avian flu, and untold numbers of deaths, birth defects, and the like may be attributed to the high levels of mercury and other toxins in the environment.

—Gregory McNamee

TOP LEFT: *The pollution over the city of Ulaanbaatar in Mongolia is exacerbated by its position in a valley and the three nearby coal-fired power stations. Photo: Barry Lewis/Corbis*

BOTTOM LEFT: *An aerial view of the new, modern housing sprawl in Shanghai beneath a thick layer of urban smog. Photo: Justin Guariglia/Getty Images*

China has 16 of the world's 20 most polluted cities, according to the World Bank. Its carbon dioxide emissions have nearly doubled since 1985, to 3.4 billion tons, with the result that temperatures have been rising regionally (as they have globally), affecting weather systems and reducing the amount of rainfall and, visibly, the annual flow of the nation's major river systems. At the same time, glaciers in China's high mountains are melting at a rapid rate, causing increased erosion and flash flooding in the environmentally sensitive uplands.

China is relatively poor in resources. But it has abundant stores of coal and can buy even more, for there are about a trillion recoverable tons available now worldwide, making it the single largest source of fossil fuel yet known. Coupled with coal's relative abundance is the fact that coal-burning technology tends to have a long life span. Many coal plants are designed to last at least 75 years, resulting in little direct economic incentive for operators to switch to other fuels or cleaner technologies. As of 2006, coal provided two-thirds of China's energy needs. Consumption is expected to double by 2025, to more than 3 billion tons a year, twice the amount used in North America, and nearly four times the amount used in Europe. By 2030, China's coal consumption could conceivably surpass that of the rest of the world combined.

Petroleum consumption has also risen dramatically in China, as a growing economy and changing social structure have made it possible for more and more consumers to purchase private automobiles. The numbers are mounting rapidly. In 2000, there were 6 million passenger cars in the nation; by 2005 that number had risen to more than 20 million. As China enters the international automotive market, exporting inexpensive passenger cars and light trucks to Europe and America, its output will increase dramatically. China's highway system is already second only to that of the United States in extent, and the Chinese government announced in 2006 that it was embarking on a program to expand the system to 53,000 miles by 2035. The U.S. interstate highway system, which turned 50 in 2006, comprises more than 46,000 miles.

Just as China is binging on petroleum and addicted to coal, the United States seems to be reversing a long tradition and using more coal itself. Fifty new coal-burning plants were approved for construction between 2003 and 2006 in the Midwest alone, in a government and business climate that seemed largely indifferent to the consequences of environmental pollution. Consumers have voiced concern for a clean environment, however, and have pressed their claims for the same in court. The result is that most of the new U.S. plants have been or will be constructed to new standards that use cleaner-burning, gasified coal or that make use of effective pollution control devices.

Because it is possible to burn coal relatively cleanly, environmentalists in the United States and Europe have suggested that industries there donate the scrubbers necessary for China to clean up its act. Coal gasification, the process whereby coal is convened into clean-burning gas, has become common, although it is initially expensive to introduce into economies that rely on abundant and much cheaper "dirty," or soft, coal.

China, in common with many other emerging economies in East Asia and elsewhere in the developing world, lacks a tradition of environmental organizations, litigants, consumer activists, and governmental agencies that, among them, keep polluters at bay. But like everything else in China, it seems, even that is changing: in 2006, when factory pollution sickened a five-year-old boy in Xinchang, a city in the coastal province of Zhejiang, his mother did some research and lodged a complaint that cancer and lung-infection rates had skyrocketed since several new chemical-producing and coal-burning plants had gone in. Even though she often worked anonymously so as to avoid trouble with corrupt officials, she helped organize a grassroots campaign that numbers some 30,000 activists across the country. In 2005, in neighboring Fujian province, more than 1,700 citizens joined a case against a polluting chemical factory, the world's largest class-action environmental lawsuit up to that time.

Things may continue on that course for some time to come, for which reason the International Energy Agency predicts that China will account for more than a fifth of the growth in world energy demand in the next 25 years, and for more than a quarter of the increase in greenhouse gas emissions. Yet, for its part, the Chinese government released a five-year economic blueprint early in 2006, announcing concern for the high environmental costs of China's economic boom and even acknowledging for the first time the fact that its rivers were polluted and its cities wreathed in smog. Said a government minister, "We all want to breathe more clean air. We do not want to pay too big a price tomorrow for growth today. If that is the case, it is not real development."

The problem is not China's alone. The costs of allowing pollution to go unchecked will in the end be far higher than for those of establishing clean-burning power plants and investing in environmental cleanup. Even if, as critics of the Kyoto protocol charge, the wealthy and developed nations of the world may end up picking up the tab for the poorer ones, the inhabitants of every corner of the world have a dear stake in assuring that China's environment suffers less damage in the near term.

—Gregory McNamee

A Vietnamese woman looks up from her search for pieces of rejected coal, washed into the river from a nearby coal mine. Workers can earn up to eight dollars a day for the coal. This work is mainly done by women, who wear hats and scarves to protect themselves from the coal dust. Photo: Barry Lewis/Corbis

POLAR WEATHER:
PAST, PRESENT, AND FUTURE

LEFT: *These hardy Adélie penguins in Commonwealth Bay, Antarctica, easily endure extreme cold and violent winds in the place that was named "The Home of the Blizzard" by early Australian explorer Sir Douglas Mawson. Here "katabatic" (gravity) winds blow for days on end as the cold dense air flows down off the Antarctic ice cap and is funneled by local terrain. The constant wind collects snow and ice particles which create what is known as spindrift. Photo: Jonathan Chester*

ABOVE: *When small icebergs erode below sea level from melting and wave action they can become unstable and roll over revealing unusual patterns in the ice. These striations can be caused by air bubbles that come out of the dense water ice as it melts and rushes to the surface along weaknesses and cracks. Small such bergs are called growlers and larger ones "bergy bits." Photo: Jonathan Chester*

At the far end of the earth, scientists continue to work as they have for decades under the most extreme conditions. Based in permanent camps in Antarctica, thousands of specialists take advantage of the unique characteristics of this continent to probe and measure the weather.

Up to 9,000 people may be involved in research work or support services, at least during the "warm" season, from October through February. With the seasons reversed from those in North America, the coldest weather runs from March through September, but even then, 1,000 or more hardy researchers remain to continue their investigations. During the cold season, the frigid temperatures prevent planes from flying, effectively cutting off live contact with the outside world. In independent and shared stations, staffs of scientists and support personnel represent research efforts from the United States, Australia, Brazil, China, France, the United Kingdom, Russia, Italy, and South Africa.

The frigid ice and snowpacks found there hold clues to changes in global weather conditions throughout long periods of history. Examining ice cores there reveals details about the variability of climate and weather over time, a range of conditions that may help scientists to interpret changes that will affect global climate in the future. When complete, the record could extend as far back as 420,000 years. At Vostok Station, two decades of drilling have produced the longest ice core ever, with a total length of 2.25 miles. Not only does this unbroken record indicate that the earth has gone through four ice ages (one roughly every 100,000 years), but analysis of air bubbles and dust particles trapped in the ancient ice produces a clear picture of what was then taking place in the atmosphere.

Although some traces of airborne pollution can be found in Antarctica, it is generally thought to have the cleanest air on earth. One of the ongoing research programs there is the Clean Air Laboratory run by NOAA, a project that tracks the amount of ozone at high altitudes above the South Pole. Another program is tracking the movements of vast ice fields that cover parts of the continent, in some cases extending out into the ocean.

Weather monitoring from surface stations is more important in Antarctica than elsewhere in the world because satellite images of weather systems are not available all the time, a situation linked to the orbit schedules of those satellites that cross over the poles. Yet satellites do provide a significant new tool in the study of Antarctic weather, yielding new kinds of data from high-resolution equipment, including thermal infrared and microwave imagers, that add to and enhance information from ground-based monitors.

In recent years, Antarctica weather study has included an increasing number of automated stations (about 65 at latest count), including buoys in the open ocean and the icepack, where harsh conditions can cut the operating times of units to less than nine months. Earlier versions of the automated stations had to be carried in on the ground and set up by hand, but the latest equipment includes units that can be deployed by helicopter. But "old-fashioned" human effort, using an array of traditional and modern technology, still provides much of the weather information, with 31 stations staffed by weather watchers. Readings involve simple surface measurements—such as temperature, barometric pressure, and wind speed—but also include many more complex factors. At some stations, balloons are used to acquire additional data from high in the atmosphere. Many factors have an effect on local and regional weather forecasting in this distant land, including rugged mountain ranges, extreme low temperatures, strong surface inversions, low water vapor content, heavy cloud cover, severe icing, and blizzards. Along with the lack of reliable long-distance telecommunications links, forecasting efforts are often difficult undertakings.

At the ice-core drilling project, future discoveries are anticipated. Just below the level at which drilling has now stopped is a large lake of fresh water, trapped beneath the ice. The water in this lake is thought to be extremely pure, uncontaminated by contact with the atmosphere for hundreds of thousands of years. The next stage of Antarctic drilling will involve extracting sediment samples from the floor of the lake; these layers of debris have gradually piled up to a depth of more than 1,640 feet and hold additional clues about conditions on earth long ago. While the climate secrets of the past emerge, data and knowledge accumulating from studying current weather patterns on this continent will help to improve computer models for current weather systems around the world.

—Kim Long

POLAR EXTREMES

The climate on the continent of Antarctica is the most extreme found on earth, too cold for plants or animals, with a few exceptions. Unlike the continent, the Antarctic Ocean has abundant wildlife, including penguins and seals. The hottest temperature ever recorded at the South Pole is eight degrees Fahrenheit; the highest temperature for the continent is 58 degrees Fahrenheit. During the "warm" season—October through February—the highest average temperature is -15 degrees Fahrenheit at the pole itself.

Average temperature extremes at the pole for July: -69 to -80 degrees Fahrenheit. Average temperature extremes for January: -17 to -22 degrees Fahrenheit. The lowest temperature ever recorded on earth is -129 degrees Fahrenheit, recorded on July 21, 1983, at Vostok Station in the east-central region of the continent. The extreme cold on this continent also contributes to a perpetual dryness; it is usually too dry to snow. In an average year at the South Pole, about nine inches of snowfall is measured, but some of this comes from windblown drifting, not fresh precipitation. In a really dry year, less than one inch of precipitation will be recorded. Other polar weather problems include high-velocity winds—more than 100 miles per hour at times—fog, and blowing snow that reduces visibility to almost zero.

—Kim Long

The Lemaire Channel between Cape Cloos on the Graham Coast (left) and Booth Island (right) on the west coast of the Antarctic Peninsula is the most spectacular navigable passage in all of Antarctica. While it is easily choked with icebergs and pack ice, most tourist expedition cruise ships attempt to make it through this narrow gap. When the sun dips down below the cirrus clouds, the reflected color and open pack ice make for a dramatic setting. Photo: Jonathan Chester

LEFT: *Cumulus clouds shroud the peaks of the remote South Orkney Islands that lie in the Scotia Sea. The strait between Coronation and Laurie Islands is usually choked with icebergs and pack ice. Remote islands like these are the breeding grounds for thousands of sea birds including cormorants, or shags as they are also known. Photo: Jonathan Chester*

RIGHT: *In spring the depth of the sea ice is measured with an ice auger offshore from Australia's Mawson Station on a regular cycle for glacial research and to make sure it is still safe for vehicular travel.*

Ice cores collected by drilling into the icecap reveal much about the climatic history of the Antarctic. Australian glaciologist, Dr. Ian Goodwin at Davis Station, measures the cores' dimensions in a sub-zero freezer lab where the cores are stored. Clearly defined seasonal layers can be identified in these cores helping to date the concentrations of carbon dioxide (and other chemical elements) in the atmosphere at the time the snow fell. Photos: Jonathan Chester

FOLLOWING PAGES 218–219: *The mountainous spine of the Antarctic Peninsula (also known as Graham Land) seen from the Bransfield Strait has many peaks rising to thousands of feet. This rugged terrain has significant effect on the local weather as can be seen from the stratus and stratocumulus cloud formations hovering over the peaks. Icebergs that break off from the ice shelves and glaciers to the south are moved north by the prevailing currents. Photo: Jonathan Chester*

FOLLOWING PAGE 220: *Sea fog in the summer is common in Antarctic coastal waters, making navigation difficult, but this will invariably burn off as the air is heated by the sun. Photo: Jonathan Chester*

FOLLOWING PAGE 221: *High-pressure systems close to the Antarctic continent in summer can bring stable conditions for days on end, but bands of high altocumulus cloud usually herald a change in the weather. Tabular icebergs like these at the mouth of the Weddell Sea move under the influence of currents but are often grounded close to the coast, making navigation a challenge, especially when pack ice is also present. Photo: Jonathan Chester*

FOLLOWING PAGES 222–223: *There is daylight around-the-clock south of the Antarctic Circle in the high southern summer. When the sun dips down below the low clouds and illuminates them from below, you can get the most dramatic sunset effects that last many hours as the sun skims the horizon and then begins to rise again. The salmon sky reflections blended with the azure glow of the water ice create a rich panorama. Photo: Jonathan Chester*

365 DAYS OF WEATHER

JANUARY

365 Days of Weather compiled by Vince Miller, meteorological data contributor to the Weather Guide Calendar.

1 1947: A hailstorm with some stones larger than cricket balls (some weighed 4 pounds) ravaged part of Sydney, Australia. Great damage to tile roofs, windowpanes, and cars; many people were injured.

2 1982: Lightning is not just a warm season phenomenon. Lightning struck two men who were hunting near Tom, OK; one of the men was killed.

3 1883: A meteor of remarkable brilliance was observed in the early evening in parts Midwest and Ohio Valley. Several observers noted its trail remained visible for 20 to 30 minutes.

4 1917: January's 5th deadliest tornado killed 16 students and hurt 10 at the Choctaw Indian Baptist Mission school in Vireton, OK. The building disintegrated as the teacher (whose jaw was broken) tried to keep the school's door closed.

5 1962: Two tornadoes, 100 yards wide and about 100 yards apart, traced parallel southeast to northwest through Crestview, FL. Minor damage occurred between the tornadoes, with almost complete destruction within the paths. One person killed, 30 hurt, 325 homes/businesses damaged/destroyed.

6 1880: Occurring before "official" records began, Seattle, WA, was buried by 48" of snow. Seattle's greatest official single snow is 32.5", which fell in 1916 from January 31 into February 2.

7 1992: Until today, the Northeast had never recorded a January tornado. Today, 6 tornadoes (5 F1's and 1 F2) struck near Grand Island. All were brief touchdowns. Several barns destroyed and roofs damaged. Temperatures were in the 40s.

8 1980 (8–11): Wind gusts over 70 mph blew out every glass pane in the Volcano House, a famous local resort. Rain (41" Haleakala Summit), and waves (to 20' on Kona Coast) caused damage of $27.5 million—Hawaii's record storm total. Six deaths.

9 1963: Fishermen readied nets and boats as huge swells broke on a nearby bar at Lagoa do Santos Andrees (Portugal); women and children watched. An enormous wave surged into the lagoon. The backwash carried 80 people into the ocean; successive breakers drowned 17.

10 1962: A vast mass of ice broke from the north peak of Nevado de Huascaran, Peru's highest mountain. The resulting avalanche of ice, rock, and debris covered 11 miles in 15 minutes; at its end it was about 1 mile wide and 15 yards deep.

11 1951: An F2 storm damages about 150 homes in Sunnyvale, California—24 were unroofed or moved from their foundations. Some businesses and the railroad depot were also destroyed. 30 people injured.

12 1997: Because of an already dreadful winter, President Clinton declared all of ND a disaster area. This was the first such U.S. statewide declaration on record. In January alone, many ND communities spent up to 10 times their annual snow removal budget.

13 1975: Canada's lowest wind chill temperature of -135°F occurred at Pelly Bay, Northwest Territory, when the air was -60°F and the wind was 35 mph.

14 1991: A nighttime tornado and electrical storm ripped across suburbs of Sydney, Australia; 1 person died and 50 were injured. Many homes were destroyed; damage was greater than $240 million. Over 30,000 homes were without power for up to 24 hours.

15 1999: Steam from paper mills in Green Bay, WI, produced clouds, which were trapped under an inversion. The moisture in the clouds had time to coalesce into snowflakes. Up to 1' of artificial snow fell on Allouez, downwind of the mills.

16 1968: An average of 42 dust storms are observed per year at Ashkhabad, Turkmenistan. But today's storm, which reduced visibilities to <6' and deposited a dust layer to 2" thick, was without precedent. Transport and water supplies were totally disrupted.

17 1950: A grass/timber fire that started near the Broadmoor Hotel (Colorado Springs, CO) was fanned by gusts up to 75 mph as it raced SE; 92 army buildings destroyed; 17 damaged at Camp Carson. Eight firefighters died battling the blaze; 20 injured.

18 1980: During a dense fog, the Alno Bridge, 30 miles north of Gothenburg, Sweden, was rammed by the 18,000 ton Liberian ship *Star Clipper*; the bridge's 500-yard-long span collapsed. Seven vehicles drove/fell off the edge of the bridge; 8 deaths.

19 1977: Snowflakes were seen as far south in Florida as Miami Beach and Homestead.

20 1993: Part of a roof was torn off a structure by strong gradient winds at Post Falls, ID; the roof was airborne for about 70 yards before hitting a pickup truck. The driver was only slightly hurt, but 3 men working on the roof section died.

21 1959: In Atlanta, lightning set fire to a bed as a woman was making it. 35 miles southwest of Atlanta, extreme turbulence caused 2 external fuel tanks to fall off an aircraft flight. No injuries in either incident.

22 2000 (22–23): A major ice storm slammed northeast, AL, north GA, and northwest SC; up to 500,000 customers still without power two days later.

23 1916 (23–24): 44°F might not seem particularly warm. But compared to the -56°F it was 24 hours later, maybe it was not so bad after all. Browning, MT's 100°F temperature change in 24 hours is the U.S. (and maybe the world) record for such an event.

24 1969 (23–24): A blizzard raged across most of MN. Winds up to 45 mph blew up to 12" of new snow and 30" of accumulated snow into huge drifts. A Jeffers man died of carbon monoxide poisoning when a snowdrift blocked his home's chimney.

25 1837: At 7 P.M. a display of the Northern Lights danced above Burlington, VT. Its light was equal to a full moon; snow and other objects reflecting the light were deeply tinged with a blood-red hue. Blue, yellow and white streamers were also noted.

26 1983 (22–29): A series of storms battered CA with wind, waves, rain, snow and record high tides. Flooding and mudslides followed. Northern CA coast registered up to 30' waves, up to 32' waves on central coast, and 16' waves at Malibu.

27 1997: At Shreveport, LA, a mother washing her child's hair in the sink during a thunderstorm received a jolt of electricity and was thrown across the room; her hand was on the child's head when lightning hit. The child was unhurt except for head tingling.

28 1887: Snowflakes "larger than milkpans" reportedly fell near Fort Keogh, MT. They measured 15" across by 8" thick, and made patches all over the fields within an area of several square miles.

29 1993: Africa's worst weather-related disaster in 1993 occurred as a result of flooding rains over Kenya's southeastern lowlands. The rain-swollen Ngai Ndeithya River collapsed a railroad bridge as a train was crossing causing 104 deaths.

30 1814 (1/31–2/4) Floes of ice on the Thames River backed up behind London Bridge (England), and then froze together. The ice was strong enough to support refreshment/entertainment stands for 5 days.

31 2006: Global warming, or just an unusual month? This January was the warmest on record for the contiguous 48 states (back to 1895). 15 states (AR, IA, IL, IN, KS, MI, MN, MO, MT, ND, NE, OH, OK, SD, WI) had their warmest-ever January.

FEBRUARY

1 1916: Seattle, WA, averages less than 10" of snow a season. Today's 21.5" snowfall was not only several average years of snow, but the city's all-time 24-hour record snowfall. Transportation in western WA was crippled.

2 1976: An explosively deepening low pressure (unofficially to 27.9" in Wiscasset, ME) caused a storm surge to come up Penobscot River; water was 12' deep in downtown Bangor in 15 minutes; 200 cars disappeared in the icy water. Heroic rescues kept death toll to 0.

3 1982: A home near Panama City, FL, was set on fire by lightning. Gas leaking from a gas line damaged by the lightning ignited; the resulting explosion destroyed the house. Amazingly, none of the 4 people in the house were hurt.

4 2006: Winds to 78 mph (produced by an offshore 968 mb 28.6" low) raked northwest WA. The Hood Canal Bridge and Evergreen Floating Bridge were closed; ferry service ceased. Seattle's Woodland Park Zoo was closed due to concern falling trees might smash fences and release animals.

5 2001: 18" of snow in FL, with snowfall rates exceeding 4" per hour. Schools and businesses closed because of the storm. It did happen, but this Florida is in MA. Parts of western MA were buried under 1 to 2' of snow. Storm totals included 23" at Savoy and 18" at Florida.

6 2006 (6–15): Extreme and severe drought conditions over most of AZ. A wildfire started by an abandoned campfire on the Mogollon Rim 12 miles north of Payson burned 4,200 acres before being contained. The same area normally has 12+" of snow this time of year.

7 1998 (4–7): Up to 4.5" rain fell on saturated soils of north/central GA (2–4); resultant flooding caused up to 20' rises on rivers/creeks. In Laurens County, a close call. 11 children evacuated safely from a stalled school bus before it floated down the Oconee River.

8 1957 (7–9): Ice storms in HI? A general 3" of clear ice collected on surfaces at the top of Mt. Haleakala, Maui (elevation: 10,000'). Three miles of power lines toppled. Wind gusts to 100 mph; coldest temperatures 24°F.

9 2006: Death Valley, CA, is the hottest (and driest) location in the U.S., having reached 134°F on July 10, 1913 (only 2°F less than the world's record high). Today's 90°F high is its earliest ≥90°F of record (records began in 1911); previous record: 90°F (2/12/1996).

10 2004 (10–11): An Alberta clipper's 50 mph winds blew 1" new snow/snow on the ground into drifts up to 20' high in northwest ND; some "whiteout" conditions. The eastbound Amtrak Empire Builder backed up 15 miles to Williston when stopped by a drift 12 to 14' high and 400 yards long.

11 1999: Today's snowfall of 57" at Tahtsa Lake West, BC, set Canada's 24-hour snow record, smashing the old record of 46.5" (1/17/1974) at Lakelse, BC (which replaced the previous record set, amazingly, on 6/29/1963, at Livingston Ranger Station, Alberta).

12 1950: Seeking shelter from a tornado by going into a ditch (or other low-lying area) should be an option of last resort. Near Sligo, LA, a pedestrian died when a car rolled into the ditch into which he had sought shelter.

13 2001 (11–13): Snow in CA's Death Valley National Park? The biggest snowstorm in many a memory of spotters and law enforcement officials left 2 to 3+' snow in some higher elevation sites. 36–42" at Towne Pass (el. 4,956') and Father Crawley Point (el. 4,000').

14 2006: An avalanche near Pass Fork of Rainy Pass of the Alaska Range swept a snowmobiler to his death; he was buried under 30' of snow. Several snowmobilers were breaking trail through heavy snows that had fallen on part of the Iditarod Sled Dog Race route.

15 2000: Amarillo, TX, set unusual temperature records today. The high of 82°F smashed the old daily high of 76°F set in 1921, and the morning low of 41°F broke the old high low record for the date of 40°F (also set in 1921).

16 2001: A car often is a poor place of shelter from a tornado; it can be lifted, rolled, bounced, crumpled, etc. A lady abandoned her car for a ditch as a tornado approached MS's Holmes County State Park; the car was blown on top of her; she died.

17 2006: A safety rule for driving in dense fog: do not stop on a freeway or heavily traveled road unless absolutely necessary. Because of fog, a truck stopped on I-295 near Jacksonville, FL; a 20-vehicle chain reaction accident resulted. One person killed; 7 injured.

18 1991: Freezing rain caused a number of traffic accidents, shutting down I-81 between Marathon and Polkville, NY. After reopening, the interstate was shut down again in only 7 minutes due to additional accidents.

19 1991: Not a tornado, just a severe thunderstorm. Winds to 100 mph and hail to 1.5" diameter pounded Floresville, TX. A 2.5-ton truck was reported to have been "pushed" ¼ mile into a field. Some children were cut by glass shards when hail broke bedroom windows.

20 1985: Lightning hurt a large tree, formed a fireball, and rolled into a house in St. George, KS. Phones started ringing all over town. The lightning strike was so bright that students at a school 2 blocks away thought the lights had been turned on and off.

21 2005: Hail (to 2.75" diameter) pummeled Canton, GA. Canton's public works department had to clear roads of ice; some streets lined by hail 15 hours after the storm. Three large car dealerships had entire inventory damaged ($3 million). 100 cars parked at a Walmart damaged.

22 1997: Rare (for the location) wintertime lightning at Oxford, NH. One bolt hit a tree and traveled to an underground cable, which exploded. A ditch 75' long and up to 4' deep resulted in the frozen ground. A different bolt struck another tree, killing a chained dog.

23 1997: Although terrain-enhanced annual rainfall can pass 400" in parts of HI, most coastal sites receive only 25 to 30". Honolulu received 2.18" in 24 hours ending at 8 P.M., about ⅒ of its average annual amount.

24 2001: In a house without a basement, an interior bathroom offers a good place of safety from most tornadoes. A two-story house near Greenwood, MS, was destroyed by an F3 tornado. A family of 4 survived in the first floor bathroom, the only part of the house not destroyed.

25 1994: 5 to 8" snow fell on southern third of lower MI; winds to 35 mph; near zero visibility. Accidents closed parts of I-94, I-96, I-75, and US-131 for hours. Near Grand Rapids, a car hit a moving train; the driver said she did not see the train crossing in front of her.

26 2004 (26–27): Snow thunderstorms produced snowfall rates of 2 to 3"per hour in parts of NC's piedmont into upstate SC. Snow 12 to 22" deep common from southern part of metro Charlotte, NC, to Rock Hill, SC; some roofs collapse.

27 2003: The 2nd heavy snow in a week led to a 14-vehicle pile-up on I-40 in Flagstaff, AZ. In the pile were four 18-wheelers and an 8,000-gallon gasoline tanker. Fortunately, no ignition occurred. However, traffic backed up for several miles and I-40 was closed for 5 hours.

28 1902: Spring flooding from Buffalo River and Cazenovia Creek in S Buffalo, NY, was characterized by waters filled with floating ice and wrecked outhouses, wooden sidewalks, boxes, and barrels. Several families were rescued from the second floors of their houses.

29 2000: This "Leap Day" brought record warmth to WI. The 59°F high at Wausau is its highest February temperature on record. Calendar-day highs were set at Rhinelander (58°F), Green Bay and Oshkosh (59°F), Madison (60°F), and Milwaukee (61°F).

MARCH

1 1910: The deadliest avalanche in U.S. history swept engines and carriage cars from 2 snowbound trains (on a grade leading to Stevens Pass, WA) into a canyon. Fatality estimates range from 96 to 118. The Wellington Station house was also swept away.

2 1960: An ice storm, worst in memory, knocked down hundreds of miles of power and telephone lines in northern Alabama; adjacent parts of northern Georgia covered by 1 to 4" ice. 90 percent of trees in Alabama's Dekalb County damaged. Ice stayed on the ground in some places until March 11th.

3 1985 (1–4): A major winter storm left about 80% of South Dakota's roads blocked from late today to the evening of the 5th, 10 to 20" of snow general; 33" at Milbank and Summit; drifts to 20'. Power poles/lines downed due to ice in extreme southeast.

4 1841: President William Henry Harrison took his oath of office on a chilly (48°F), cloudy, windy day. His speech took 100 minutes; he rode a horse to and from the Capitol without a hat or overcoat. He caught a cold, which went to pneumonia, and he died one month later.

5 2004: Strong gradient winds (to 59 mph) caused much damage in central Illinois; several semis blown over. In Paris, an SUV driver died when a falling tree limb smashed through the roof of her car. A Champaign, IL city bus was blown off a bridge and fell 15' into a creek; 6 hurt.

6 2005: An avalanche near Colorado's Aspen Highlands Ski Area killed a man involved in an avalanche awareness class. While skiing across a slope, he fell, rolled, and triggered the avalanche, which carried him almost all the way down the mountainside.

7 1960: Dust devils are not just a western U.S. phenomena. A dust devil in Sudbury, MA, ripped a small plane from its moorings, sent it flying 40' through the air, and smashed it against a fence.

8 1990: Near Ward, AR, 41 students and their driver were stranded in their school bus in water 4' deep; flooding had washed out bridges. After 2½ hours, they were rescued without any problems.

9 1998: Good thing this morning's F1 tornado, which destroyed 20 amusement park rides, along with bleachers and outbuildings, did not hit the St. Lucie County Fairgrounds (near Fort Pierce, FL) when the fair was in session. $3.2 million damage.

10 1986: Near Livermore, KY, during a severe thunderstorm, a 58-year-old man tried to hold a barn roof down with a chain. The winds of the storm prevailed; the roof was blown off and the man was thrown 78' in the air. He died of chest and head injuries.

11 1956 (10–11): Winds to 80 mph in southern lower MI caused much damage. Blizzard conditions in northern MI; snowdrifts 6 to 8' common. Many northern communities isolated 1 to 2 days. Snow in the north accompanied by lightning and thunder, somewhat unusual for this time of year.

12 1993 (12–13): "Storm of the century" mauled Florida—tornadoes, hail, wind, snow (to 4" north of Pensacola to near Crestview) and an unprecedented winter storm surge along the Gulf Coast (to 12' in Taylor County; 10 deaths). 25 direct deaths, $1.6 billion damage.

13 1993 (12–13): The worst winter storm in Alabama history left snow over the entire state; 12 to 20" Birmingham to northeast corner of state; 2 to 4" common along the Gulf Coast. Frequent lightning at times gave the atmosphere an "eerie blue-tinged glow." 400,000 homes without power.

14 1990: Talk about a quick reaction. Near Halley, AR, a couple heard a loud "pop" during stormy weather. They immediately went into a nearby closet for shelter. Good thing, because an F1 tornado destroyed three-quarters of their house. They were uninjured.

15 1998: A "rogue" wave swept out of the Pacific Ocean into the harbor parking lot at Port Arena, CA. Thirty cars were displaced and much sand and debris as well as many logs were deposited in the lot.

16 1965: Severe storms wracked parts of OK's Grant and Kay counties. Scattered heavy hail fell for 5 to 20 minutes along most of the path. At Nash, hail ranged in size from ¾ to 2" (ground covered to 2" deep), with some fist-sized clusters containing 5 stones each.

17 1894: With today's 80°F high in Sioux Falls, SD (a record for the date, and until 3/7/2000, the earliest 80°F on record), one might have thought winter was leaving. But –4°F record lows on the 25th will show the fickleness of Great Plains weather.

18 1925: The Tri-State tornado became the deadliest (695 killed; 234 at Murphysboro, IL) and longest (219-mile continuous path from near Ellington, MO, to northeast of Princeton, IN) tornado in U.S. history. 66 school deaths—including 33 at Desoto, IL (U.S. record).

19 1963: At Becks Mill, IN, 2 women died when their shelter from a tornado collapsed on them. The fruit cellar (behind the house) may have been weakened by rain; the house suffered only broken windows. One of the women had been hurt by an F3 tornado on 4/3/1956.

20 2003: In Georgia, there's seldom enough snow to get out of school, but flooding, now that is something else. Early A.M. rain closed 30 roads (washed out 10) in Laurens County. County schools let out early because of extensive flooding and road closures countywide.

21 1932: Alabama's deadliest single tornado (an F4) killed 49 people on its 60-mph path from north of Marion to the northwest corner of Coosa County. Entire families were killed, and 150 families were left "destitute." four other tornadoes killed 30 or more the state.

22 1897: Two teachers and a number of students of Arlington (GA) Academy stood at windows observing a storm. Suddenly, an F2 tornado destroyed most of the building. At least 8 of the school's 29 students died.

23 1990: High winds and high lake levels caused an ice breakup on Lake Champlain. Wind-driven blocks of ice caused much damage to docks, boats, and lakefront cottages on NY's and VT's shorelines. Locals said the unusually clear ice made it especially destructive.

24 1990: Dime-sized hail fell on Union City, OK, during a thunderstorm. Not unusual, except the temperature at Union City was almost freezing. In fact, the same storm also produced sleet and freezing rain.

25 2000: Near Shelter Cove, CA, a rogue wave swept a lady with a Canadian school group into the ocean. Four members of the group tried to rescue her, but were overcome by waves and currents. Two of the rescuers were rescued by a fishing vessel and the Coast Guard.

26 1999: South winds to 50 mph roared most of the day across central/north-central SD. Blowing dust blocked out the sun and reduced visibilities to near zero at times. In Selby, part of a hotel's 2nd- and 3rd-floor front wall was blown off by the wind.

27 1961: Wind gusts across Illinois, often 50 to 80 mph; 2 killed, 19 injured. A 100 mph gust toppled wind equipment from roof of the FAA station at Joliet, IL. A man died when blown off a steel beam at Streator Township High School; a Mason City TV repairman hurt when blown off a roof.

28 1956 (27–28): Gusty winds to 80 mph damaged $1.5 million acres in eastern Colorado; pilots noted dust to 20,000' above ground. Dust closed roads east of Limon due to near zero visibility. Schools at Arriba, Bennett, Stratton closed due to dust and wind.

29 1992 (28–29): Heavy rains falling on a deep snowpack caused ice-jam flooding at Blanchard, ME (Piscataquis County). Chunks of ice 50" thick and 10 to 15' across moved a barn 75 yd. Tractors and pickup trucks also found new resting places thanks to the force of the ice.

30 1998: Severe thunderstorms and heavy rains raked southeast Wisconsin today. At Lakeland College (near Johnson), the campus wastewater treatment buildings were flooded and sewage backup problems ensued. Classes were cancelled the next day.

31 1959: Hail to 11.5" in circumference dug holes in the ground south of Thackerville, OK; the stones also came through the roofs and broke neon signs. At Shawnee, large hail knocked a 17-year-old boy unconscious.

APRIL

1 1978: In Latrobe, PA, a gust of wind picked up an elderly man and hurled him into the sidewalk; he was killed. Winds at the Latrobe airport were measured at 85 mph.

2 1956: As a tornado passed through parts of Wilmette, IL, in the early morning, its barometric pressure reductions resulted in "rear windows of cars blown out" and "sealed beam headlights popped out of their receptacles."

3 1956: Severe storms (wind, rain, hail, and a few tornadoes) swept across much of IL. At Joliet, peak winds gusted to 109 mph before the anemometer was destroyed. In Chicago, 9 people were injured when they were blown through a barricade and into an excavation.

4 1968: Would homeowner's insurance have covered this? Winds to 50 mph pushed Lake Champlain ice up and over embankments at Tabor's Point (West Swanton, VT). Ice piles to 30' high approached/threatened 10 houses. In one case, the ice finally stopped 10' from a house.

5 1951: 120 people were lying on the floor of the Bridge Creek School near Blanchard, OK, when it was unroofed and its top part destroyed by today's F2 storm. Only 2 injuries. Better luck than on 5/3/1999 when 12 people were killed in Bridge Creek area by an F5 tornado.

6 1999: The chimney of the Mother Irene Gill Library (College of New Rochelle in New Rochelle, NY) was hit by lightning. Stone/concrete crashed into the hall below, occupied by 20 faculty/students. They escaped uninjured.

7 1988: 60+ mph winds caused a dust storm that deposited salt on power lines in southeast Idaho; rain dissolved the dust. Chemical reactions ignited material in the insulators, thus causing many power pole fires. Power outages ensued for 7,000+ customers.

8 1998: Lightning hit a tree and shed at Bellevue, WA. It knocked a hole in the roof of the shed and caused 2 bags of lawn chemicals stored in the shed to explode.

9 1991: A squall line of severe thunderstorms (gusts generally to 60+ mph; some to 110 mph) raced across WV; serious damage in 49 of 55 WV counties. two deaths, 86 injured, $16 million damage. Some rural counties had nearly all roads blocked by fallen trees.

10 1997 (10–11): 6 to 14" rain fell on TX's Lavaca County. A van washed off a road between Yoakum and Sweet Home; 2 couples followed a fence line to a tree. They clung to the tree for 2 hours, having to endure both rising water and fire ants. They survived both ordeals.

11 1956 (10–11): A dust storm raged across Oregon into southeast Washington before rain falling through the dust caused a "brown rain." A muddy mess coated sidewalks/buildings/cars. Driving impossible at times—as wipers could not remove mud.

12 1934 (11–12): Today's 121 mph gust atop Mt. Washington, NH (el. 6,288'), became the world's highest recorded surface wind. A 5-minute wind speed of 188 mph was also recorded today. The 24-hour average was 128 mph.

13 1987: In Memphis, TN, a man was struck by lightning while pumping gas into his car. Several hours earlier, 2 men were hit by lightning while working on a parked aircraft at the International Airport. Fortunately, all 3 survived.

14 1988: Lightning struck a tree near a home in Moab, UT, and came into the house on the underground phone lines. A 13-year-old girl talking on the phone was seriously burned. She was lucky; several people are killed in the U.S. each year in similar circumstances.

15 1982: The earliest-in-the-year tornado of record in ND was on the ground for 6 miles near Lisbon. The F2 storm was up to 440 yards wide. It destroyed a barn, uprooted trees, and drove tree branches through steel granaries.

16 2003: Just 1 lightning bolt. It hit an ONCOR transmission line near Granbury, TX, and resulted in 4 power plants temporarily shutting down. Up to 400,000 customers were without power for up to 2 hours.

17 1994: In Fort Fairfield, ME, 3'-thick chunks of ice littered lawns due to earlier flooding of the icy Aroostock River.

18 1950: A small house was thrown 200 yards into another home by an F2 tornado near Mobile, AL. As the storm passed near the University of South Alabama campus, a seismograph recorded a 73-second-long vibration produced by the passing tornado.

19 1968: A devastating hailstorm (stones to 2.5" diameter) raked the Long and Short communities in OK's Sequoyah County. Southeast of Short, hail stripped bark off the windward side of cedar trees. Cattle were "literally beaten to death" by the hail.

20 1896: For some readers, 100+ year-old descriptions may be a bit hard to visualize. An F4 tornado killed 2 (in a totally flattened home) and hurt 20 at Booktown, OH. The tornado "crept along the ground like a mammoth rock" and "smoke puffed at the top like an engine."

21 1951: An F4 tornado tracked 30 miles from near Lebanon to near Soso, MS. Two people died when a house was leveled. A body was found 200 yards from the site, and a refrigerator 300 yards from the home.

22 2001: A lightning bolt struck a house in Sioux Falls, SD, and injured a person standing in the basement. Strange, but here's why: the basement was "wet," so the charge of the bolt was able to find a pathway via the moisture to the victim.

23 2001 (22–23): Extreme changes. Pierre, SD, set its heaviest snow of record for so late in the season (12.5"). By the 28th, the snow was melted and a record date high (96') was set. On only one earlier date in the year has Pierre seen a high ≥96°F: 98°F on 4/21/1980.

24 2002: Near Popular Bluff, MO, a woman took shelter from an F4 tornado in her bathtub. Both were blown 200' into the median of US-67. She survived. A big chunk of asphalt from the same highway was blown through the window of a car, hurting a person inside.

25 1910: An incredible day in the southeast part of the U.S. with a trace of snow noted as far south as Pensacola, FL. In Nashville, TN, today's 1.5"-snow was its April 24-hour record, and its latest measurable snow. Today's 32°F was also Nashville's latest freeze of record.

26 1998: To increase the odds of surviving a tornado in a home without a basement, seek shelter on the ground floor, in a small interior room, like a closet. Two people were unhurt when their home near Hale Center, TX, was destroyed by an F2 tornado—by doing just that.

27 1968: Lightning hit a pine tree near a house in Enon, MS. It ran underground, then entered the house, destroyed part of the living room, and sent wood splinters flying. A woman sitting in a corner had 2 splinters embedded in her temple, but was not seriously hurt.

28 1950: The power of violent tornadoes. An F4 tornado near Clyde, TX, blew a small refrigerator more than ½ mile and "lodged" it atop a telephone pole. Near Holdenville, OK, a couple was found (dead) in the wreckage of their home, which had been blown 150 yards.

29 1968: A dust devil with a loud "jet plane" roar damaged trailers, carports, and TV antennas in Pepperell, MA; plywood sheets "sailed away." In Franklin, a dust devil ("2 telephone poles high" and with a loud whistling noise) lifted a 500 lb. boat into a tree.

30 1898: A violent F4 tornado struck near Hospers, IA, as a father was about to go into the outside storm cellar; he had a young child under each arm. The home was annihilated. The children were found dead, still in their father's arms.

MAY

1 1929: "Lightning does not strike twice" and "tornadoes do not hit the same location twice" are both myths. The city pump house at Fort Smith, AR, was damaged by a tornado today. On 12/17/1929, it will be destroyed by another tornado.

2 1990 (2–3): Record and near-record late heavy snow fell on southeast CO and northeast NM. 18.6" near Kim, CO; 18 to 22" in NM in Des Moines/Capulin/Folsom area. In NM's Union and Colfax counties 1800 newborn calves ($1 million value) died of hypothermia and exhaustion.

3 1984: Flying debris is a potential hazard of severe thunderstorm winds. A fiberglass bench was blown off the 21st floor of Peachtree Summit Building in downtown Atlanta, GA. Unfortunately, the bench landed on a car on I-75/85, killing a backseat passenger.

4 1963: Hail size often references local features. Hail to "hedge apple" size left the ground white in parts of Coffey/Franklin counties (KS). A few larger conglomerate stones near Burlington, Sharpe, and Waverly had a 4.5" diameter and 14" circumference.

5 1998: An F1 tornado made a brief 3-mile track at Los Altos, CA. During its brief lifetime, it picked up and hurled (and injured) a tennis coach about 20' at Los Altos High School.

6 1983: 60 mph wind gusts caused dust storms (most dust from plowed fields in central IL) in northern IL; visibility near zero at times. Many accidents, including a 9-car/2 semi-trailer incident on I-57 near Rantoul. Chicago's skyline was nearly invisible from a short distance away.

7 2004: Thunderstorm winds to 83 mph overturned a number of portable toilets near the re-enactment main stage at VA's Battle of Spotsylvania. 3 Amtrak trains were stalled near Chatham by downed trees/power lines.

8 1998: Not a tornado, but the winds were stronger than most tornadic wind speeds. Downburst winds to 120 mph caused $1.5 million damage in Kingsport, TN. 100 businesses and 70 homes damaged; roofs blown off and windows blown out. Many trees and power lines downed.

9 1980: A blinding squall, then near zero visibility in fog at the Sunshine Skyway Bridge over FL's Tampa Bay. A ship hit a bridge piling, causing a 1,200' section to collapse into the bay. Several vehicles, including a bus, drove over the edge of the span; 35 deaths.

10 2000 (4–31): A prescribed burn (4th) in NM's Bandilier National Monument became a firestorm today that swept into W Los Alamos and the National Laboratory; µ250 homes/buildings wrecked. 48,000 acres burned; 21,000 people evacuated. $1.5 billion property damage.

11 1963: Not typical of fire damage found in most parts of the country. Prairie fires started by lightning and rapidly spread by strong winds burned 5,000 acres of grass in W Clark County (KS); 4 miles of fence posts destroyed at one ranch.

12 2004 (11–12): A very late-season snow/ice storm hit north-central/northeast ND. Snows 8 to 12" toppled 150 power poles in Mountrail County. High winds/ice to 1" diameter downed 750 power poles, mainly east of ND-1 from Fairdale to Canadian border.

13 2002 (12–13): Flash flooding occurred on almost every watercourse in MO's Iron County; $5.5 million damage. Many people trapped in cars by floodwater. Near Ironton, a man crossing Stouts Creek by foot to rescue his dogs was swept away or drowned.

14 1956: A few miles N of Martinsville, VA, on VA-108, thunderstorm winds tore off part of a truck's tailgate. Unfortunately, it smashed through the windshield of another car, killing the 2 women inside.

15 1834 (12–16): A wintry period from the Great Lakes to parts of New England. 6" snow at Erie, PA (14); 12" at Rutland, VT, by P.M. today. Newbury, VT, received 2' (sleighs on roads on 16th), and the Haverhill, NH, vicinity had 2' in valleys and 3' on hills.

16 1962: At the wrong place at the wrong time. 6" rain in 45 minutes caused a wall of water 6' high, which swept a pickup off a highway near Clarendon, TX; the 2 occupants drowned. The truck was swept 100' from the highway; 1 body was carried 5 miles.

17 1962: Near Pecos, TX, thunderstorm winds picked up 8" x 16" cinder blocks, blowing 1 through a car window, and others through cinder block walls. In Pecos, falling hailstones (to 3" diameter) "looked like huge snowballs" and bounced 5 to 10' upon hitting the ground.

18 1997: A thunderstorm downburst (gust to 81 mph) hit the Parkersburg, WV, airport; windows broken in 14 vehicles. The tower was briefly evacuated. The weather observer said, "We thought we were having hail, but it wasn't hail. It was busted glass flying around."

19 1973 (18–19): Parts of central NY received their heaviest, latest snow of record; 10" at Old Forge, 8" at Waterville, 6" at Boonville and Oneonta. Many trees and power lines down due to heavy, wet snow. Some schools closed.

20 1963: Thunderstorms killed 4 people in ME. A fishing boat on Bowler Pond (Palermo) was capsized by high winds; its occupant drowned. A plane crashed at Bath, killing the 3 occupants. The crash was attributed to either turbulence or a lightning strike.

21 2004: Hail to 3" deep (drifts to 10") caused a car on I-94 east of Glendive, MT, to skid off the road and blow a tire; nobody hurt. South of Glendive, a flash flood washed out a hole 10 to12' wide by 15 to 20' deep on a gravel road. Fortunately, nobody drove into it.

22 1951 (21–22): 12" rain led to disastrous flooding (6 deaths) at Hayes, KS; water to 20' deep in parts of town. 75 city blocks and Fort Hayes State College campus inundated. Now I-70 travelers are probably unaware of a flash flood potential as they race past Hayes.

23 1960: Sea waves and accompanying tidal action from Chilean earthquakes affected AK's Gulf of Alaska coast from Prince of Wales Island to Montague Island for up to a week beginning today. Today's 2 tidal waves near Yakutat were up to 14' high.

24 1951: Northeast gales caused waves to 15' high in the harbor at Newport, RI. A 50' Navy launch with about 142 men on board capsized; 19 of the men drowned in the incident.

25 1956: Dust devils swirled over eastern MA. A child and bicycle at Ipswich were tossed 40'; at Dudley, a child and tricycle were tossed 10'; both children were injured. One of the dust devils was said to be preceded by a sound "like the baritone siren of a freight train."

26 1978 (26–27): 10" rain fell in 90 minutes just west of Canyon, TX. Flash flooding destroyed 15 houses, 27 mobile homes, 300 vehicles, and 12 camping trailers as far downstream as Palo Duro Canyon State Park (swept by a 12' high wall of water). Four killed, 15 injured.

27 1963: Dense fog and vehicles traveling at high speeds make a deadly mix. At 3 A.M., a number of vehicles drove into a fog bank covering part of the NJ Turnpike at Elizabeth. Five drivers and 1 passenger killed; 10 injured.

28 1990: As the beach patrol warned people (thousands present) to get out of the water and off the beach at Wend of TX's Galveston Island, a woman was struck and killed by lightning as she attempted to get her children out of the water.

29 1959: A man fled his car and jumped into a ditch east of Washington, KS, as a tornado approached. The car was picked up and blown over him; he grabbed the rear bumper but was unable to hold onto it. Lucky for him: the biggest piece of his Rambler found was 3' x 7'.

30 2001: At Prichard, AL, while sitting on a barber's chair and talking on the phone, a man was hurt by a lightning strike. Knocked unconscious, he awoke 45 minutes later in the hospital. Other than soreness, he had no apparent serious injuries.

31 2002: A searing end to May brought record May monthly heat to Safford, AZ (109°F); McCook, NE (106°F); Goodland, KS (104°F); and Rifle, CO (99°F). The 102°F May record high in Delta, UT, followed Delta's all-time May low of 19°F set on the 9th.

JUNE

1 1956: Not good to be close to 5,000 lb. of dynamite when it explodes. Lightning "prematurely" exploded that much dynamite at a MA turnpike construction site near Woronoco. 25 men were within 300' of the blast epicenter. five were hurt; amazingly, no deaths.

2 1993: A Little League baseball game in North Salt Lake, UT, was interrupted by a tornado. Players and spectators took shelter just before the tornado hit. Many people covered with mud/debris; 2 hurt from flying debris. Windows blown out of several vehicles.

3 1860: A massive tornado swept an 80-mile path from east Cedar County (IA) to near Whiteside, IL; 92 killed (41 in Comanche, IA). 23 killed on a raft as the tornado crossed the Mississippi River; 3 survivors ended on the IL shore with no memory of how they got there.

4 1956: Dust devils caused problems in Benson, AZ, several times this month. Today, a large dust devil damaged a number of roofs in town, and demolished a small house and carport. On the 26th, a dust devil collapsed the rear wall of an unoccupied church.

5 1960: Fifteen milk cows were found dead, lying in a circle around a small apple tree near Pownal, VT. Lightning killed them, although the tree exhibited no lightning damage.

6 2002: The parking lot of a closed service station at Somers Point, NJ, was hit by lightning; it traveled to the underground storage tank. The resulting explosion left a crater 50' in diameter and 8 to 10' deep.

7 1946: Only 5 people have been killed by tornadoes in MT since 1880. An F3 tornado near Froid destroyed a 5-room farmhouse; furniture was blown over 1 mile. One fatality was due to a falling chimney.

8 1960: "Chunks" of hail up to the size of hen eggs fell for 45 minutes at Hooker, OK; rain to almost 4" accompanied the storm. Much property and crop damage.

9 2005: Tropical Storm Arlene formed at 0600 UTC (2 A.M. EDT) about 175 miles west/southwest of Grand Cayman and became the 1st of what would be a record 27 named storms for the 2005 Atlantic hurricane season. Landfall west of Pensacola, FL, on the 11th with 60 mph sustained winds.

10 2005: Unlike Hurricane Katrina, which would cause 1300+ deaths in the U.S., Tropical Storm Arlene caused very little damage and 1 death. The fatality occurred when a Russian exchange student was caught in a rip current off Miami Beach, FL, produced by Arlene's winds.

11 1965: Heavy rains fell on an unusually deep snowpack in the high Uintas of UT. After midnight, a flash flood roared down Sheep Creek Canyon, destroying 3 recreation areas, 7 bridges, and 5 miles of newly paved roads. 7 swept away while sleeping at Palisades Campground.

12 1993: A tornado that started as a waterspout over the Missouri River and destroyed a house moved onshore near Pierre, SD. The owner hid in the fireplace and survived. He was lucky, because falling chimney stone has been the cause of a number of deaths.

13 1973: At Jennison, MI, a dust devil picked up a rug, stepladder, and a patio umbrella. It also lifted a 350-lb. concrete patio table and dropped it 3' away (in several pieces after the landing).

14 1998: Iowa's 24-hour rainfall record set at Atlantic, IA, with 13.18"; resultant flooding destroyed 21 homes. Saturated soil around Lake Panorama began sliding into the lake; 3 homes "cracked" beyond repair as earth sank 12 to 18".

15 1992: A legendary storm for chasers: ≥12 tornadoes in 3 hours in Mitchell County, KS. A farmer south of Cawker City reported new tornado damage to his farm each of 5 times he went to, and than came out of, his basement. One time he noted 3 on the ground and 4 in the air.

16 1998: Today's 103°F in Midland, TX, was the 1st of 14 days in a row the high was ≥100°F. The old record: 12 days (July 14–25, 1981; June 30–July 11, 1964). This June was the second warmest of record, behind June 1990.

17 1992: Weather sometimes has no respect for majesty. Severe thunderstorm winds blew down IL's largest tree, which was also the largest eastern cottonwood in the U.S. The tree, 138' tall, was in Gebhard Woods State Park.

18 2002 (17–18): Deadly floods ravaged parts of southern Russia between the Caspian and Black seas; Chechnya, Krasnodar, and Stavropol particularly hard hit. 40,000+ homes flooded and 200+ bridges damaged or destroyed. 100+ deaths.

19 1992: Midday severe thunderstorms produced $600 million damage to Wichita and south-cental KS. At least 70 people treated for minor injuries, mainly due to hail and hail-broken glass. 377,000 acres of crops damaged/destroyed.

20 1988: At Raleigh, NC, an employee of a fast-food restaurant was injured by a lightning bolt while serving food from the drive-up window.

21 2006: The National Weather Service Forecast Office in Birmingham, AL, reported thunderstorm winds blew over 2 portapotties at the intersection of AL-49/AL-14 in Tallapoosa County. Apparently, they were not in use at the time as no injuries were reported.

22 2006: Thunderstorm outflow winds 50 to 70 mph raked US-62/82 in northeast Terry County (TX). Near zero visibility in blinding dust and sand led to multiple accidents; 15 injured. An elderly man died when he drove into a tow truck assisting at an accident.

23 2001: Neither snow nor rain nor heat nor gloom of night stays these couriers from the swift completion of their appointed rounds. How about lightning? It hurt a mailman ready to check a mailbox in Venice, FL. Rain started; his arm hair stood up just before he was hit.

24 1979: Too bad this tornado was not videotaped. Forming near the east side of CO's Pikes Peak, an F2 tornado moved down Williams Canyon into Manitou Springs. Convenience store and gas station demolished; several homes severely damaged. One person hurt.

25 1950: A Northwest Airlines DC-4 flying at 3,500' in an area of thunderstorms crashed into Lake Michigan about 20 mi. from Benton Harbor; all 58 on board died. No official cause for crash, but weather suspected.

26 1989: Near Junction City, KS, a woman was struck by lightning when she got out of her car to film an approaching storm. The bolt hit the top of her head, killing her instantly.

27 2004: A number of people took shelter from an approaching storm by standing under trees lining a parkway connecting a parking lot to a beach at GA's Buford Dam State Park. Lightning strikes hit a tree(s) and killed 3 people instantly; 6 others hurt.

28 2005: Tropical Storm Bret formed <50 miles east/southeast of Veracruz, Mexico. Combined with Arlene, this June became the 13th time since 1851 that 2 June tropical storms formed. Landfall on the 29th near Tuxpan; flooding in state of Veracruz killed at least 1.

29 1989: An F1 tornado near LaCenter, WA, lifted a car 6' into the air and turned it around 180° before setting it back down. The driver was slightly cut and bruised (and probably a bit surprised). Several trees were also blown down by this 15-yard-wide tornado.

30 1972: Lightning struck and killed a 23-year-old man as he was walking through Bay-Brook Park in Baltimore, MD. A witness said the man "flew through the air and seemed to explode." His clothes were ripped off; some pieces of clothing were found 80' away.

JULY

1 1997: 4" of rain fell in 90 minutes on a 4-square-mile area neat Pitcairn, PA; led to flash flooding. Water from Dirty Camp Run flowed through the elementary school's lower-level windows. 13 buildings swept off their foundations. $10 million damage.

2 1977: Occurring during one of the worst drought seasons in CA history, showers generated by an unusual July occluded frontal passage left .35" rain on San Francisco (CA) International Airport. This is the highest daily July rainfall on record for the airport.

3 1995: On Lake Shawnee (KS), as a woman was pulling an inner tube connected by a rope into a boat, lightning hit water nearby. She was electrocuted; 3 other people in the boat were unhurt.

4 2002: Lightning struck a vendor display area just east of the tunnel entrance to the infield of FL's Daytona Beach International Speedway. Six people were treated and released at the scene; 2 others were hospitalized.

5 1966: It would have made an interesting video segment in a television sportscast. Near Anoka, MN, a tornado "chased" 30 people from an archery range. Fortunately, they ran south and the tornado veered north.

6 2002: The Chediski-Rodeo Complex fire was contained in AZ's Navajo County. The merger of 2 human-caused wildfires (begun in June), 468,638 acres burned, as did 426 structures/homes. 30,000 people evacuated. Flame heights to 300 to 400' noted during some uphill runs.

7 2000: A farmer's worst nightmare. Hail to ⅞" diameter caused a swath of destruction in a 2-mile-wide, 20-mile-long path from west of River Falls to north of Hager City, WI. $4.5 million damage to corn, soybeans, alfalfa, and small greens. 21 square miles a total loss.

8 2000: Rash flooding occurred in Chapman Gulch near Ophir, CO, trapping a motorist who had driven through the gulch moments earlier with water up to the hood of the vehicle. The motorist was able to leave the morning of the 9th after the waters receded.

9 1995: Electrocutions have resulted from similar events. At Bristol, FL, lightning hit a tree, traveled into a nearby septic tank, and followed the piping into the bathroom toilet of a nearby house. A man sitting on the "throne" went airborne, but did survive.

10 1982: A rare event for AK: 2 funnel clouds were observed 10 miles northeast of North Pole, AK. They lasted about 5 minutes and reached to within 500' of the ground.

11 1951 (10–11): Snow fell at some locations in the northern Rocky Mountains and, incredibly, in the Black Hills. On the 10th, 6. 5" reported at the E entrance of WY's Yellowstone NP and 2" at Mystic Lake, MT. On the 11th: Lakeview, MT, 2", and Bixby Dam, SD, 0.4".

12 2001: A nearly stationary supercell thunderstorm dropped large hail and 3 to 6" rain in 1 hour southeast of Cohagen, MT; hail drifts to 10' deep were measured in coulees. Four days later, golf-ball-sized hail was still embedded in mud/straw, along with drifts to 4' deep.

13 2004: Driven by 60 mph winds, 2.75"-diameter hail damaged most buildings/vehicles in Posen, MI. Holes punched in siding and roofs. 300 holes in a church roof; a greenhouse lost 1,000 2' x 2' windowpanes. One man's back badly bruised as he moved his vehicle to shelter.

14 1987: Ball lightning hit a home in NY's Oneida County, went through the house, down the stairs to the cellar, into an oil furnace, through a repairman, and into the ground. He collapsed but survived. When he got home, it was without power due to lightning.

15 2000: Lightning is a major danger to hikers in the CO Rockies in summer. Near Conundrum Hot Springs (Pitkin County), a bolt struck 1 of a group of 4 hikers. Her clothing was shredded and boots blown off. Injuries included bleeding from nose and ears; burns on chest and feet.

16 2005: Hurricane Emily briefly a Category 5 hurricane 100 miles southwest of Jamaica; lowest pressure was 929 mb/highest sustained winds 160 mph. Emily is the only known July Category 5 hurricane of record in the Atlantic basin, and the earliest Category 5 on record.

17 2000: Intense storms produced a vivid lightning display in the early morning over Kansas City, KS/MO. 2,000 plus cloud-to-ground bolts hit Johnson County between 2 and 3 A.M.; the number of intra-cloud bolts may have exceeded 20,000 in the same time span.

18 1910: A chilly morning in the Ozark Mountain valleys of AR. In fact, this morning's low of 41°F in Harrison set the state's all-time record low for July.

19 1956: 70 tourists were at the bottom of CO's Royal Gorge enjoying the view. A 10' high wall of water from a cloudburst over Telephone Gulch roared into the gorge and knocked a small girl off the platform. A tourist managed to grab her and kept her from being swept away.

20 1998: Heavy rain from severe thunderstorms in Las Vegas, NV, combined with a clogged drainage system to cause the roof of the Palace Station Hotel and Casino to collapse. The same structure caught fire when hit by lightning several hours later.

21 1985: Near Blanding, UT, 2 women were running to their truck. One of them was jumping over a ditch just as lightning struck; she was knocked unconscious, but survived. The other died of cardiac arrest.

22 2002: Heavy rain and hail caused a land/rock slide, which covered CR-306 (Cottonwood Creek drainage near Buena Vista, CO) with debris to 10' deep/1-mile long. An elderly couple's van was swept off the road; mud/rocks flowed into it. They suffered hypothermia, cuts, and bruises.

23 2005: An Atlantic hurricane hitting southern CA? Sort of. Remnants of former Hurricane Emily (tropical moisture transported west/northwest), in the form of early A.M. thunderstorms, awoke many people from the deserts to the coast of CA's San Diego County.

24 1999: Today's high of 97°F a record for NYC, NY's Central Park (4 daily records and 1 tie by month's end; 101°F on 5th and 6th were daily records and hottest this month). This month had an average daily temperature of 81.4°F, eclipsing 1955's record 80.9°F.

25 2004: Rain to 9" caused an 18' high wall of water, which raced down Dry Creek 25 miles north/northeast of Dryden, TJC 100'-long pavement slabs of TX-349 washed away; mud/gravel to 4' high topped other highway areas. 13 workers were trapped 36 hours at a well site encircled by water.

26 1995: A prime rule for lightning safety is not to seek shelter under trees or other tall objects. Nonetheless, 9 campers at Camp Tuscaror, a Boy Scout Camp in NY's eastern Broome County, were injured when struck by lightning as they stood under tall pines.

27 1987: An F3 tornado tracked 30 miles in a general southwest direction (most travel NE) from near Red Wing, MN, to north of Fairbault. It leveled 2 farms in Welsh Township; debris/equipment was blown hundreds of yards. A horse was seen flying "horizontally" through the air.

28 1996: Only 15 to 20 yards wide, an F1 tornado blew out all windows on the south side of a house near Mercer, ND, as it passed within 20 yards. A pickup was moved 7', and the family car was picked up and spun around 180°. Shingles tom off house roof; $75,000 damage.

29 1997: A lightning safety rule is to stay away from windows. At Aspen Park, CO, a woman received minor injuries when she was struck by lightning after it passed through an office window. She was temporarily blinded for 15 minutes.

30 2001: At 12:30 P.M. CDT, thunderstorm downburst winds to 86 mph damaged buildings at the airport, bent a flagpole, and knocked down trees and power poles at Wagner, SD. An accompanying heat burst raised the temperature 20°F (to 99°F) in a few minutes.

31 1969: At Sanford, MI, a couple tried to hold the roof onto their patio when the wind threatened to remove it. As the roof did start to lift, the man let go. The wife held on and was carried over the house and landed 75' away. She died.

AUGUST

1 2002: A 76-year-old man was found dead in his uncooled apartment at De Soto, LA. According to the coroner, the apartment had air conditioning, but the man was not using it in order to save money to buy medication for his diabetic/coronary conditions.

2 2002: Thunderstorm wind gusts to 83+ mph toppled thousands of trees, power lines, and poles in NJ's Monmouth County; 168,000 homes/businesses without power. 250-year-old trees to 8' diameter downed. $1.5 million damage to Monmouth University; its offices closed 5 days.

3 1990: A record heat wave smothered England and Wales; the high of 99°F at Cheltenham the highest-ever temperature on record for the UK (until 101°F at Brogdale on 8/10/2003). At Liverpool (England) the entire stock of a chocolate factory melted.

4 1992: Thunderstorm microburst winds to 100 mph caused more than $500,000 damage to a farm at East Georgia, VT; several cows had to be killed. In 1907, a barn on the same farm was destroyed by thunderstorm winds; 42 cows died.

5 2002: Some good luck and timing for the crew. Near Glenwood Springs, CO, a freight train was halted by 4 to 5' of mud left by a flash flood. As the crew scrutinized the situation, another flash flood deposited several feet of mud and debris on the tracks behind them.

6 1961: Thunderstorm winds to 100 mph raked parts of Lake Texhoma, boundary between OK and TX. 100+ boats were sunk or damaged; similar destruction to piers. One drowned when a boat was capsized by the storm.

7 1979: Seiches (sudden changes in water levels) noted at Lake Ontario shore locations in Rochester, NY, vicinity; rises to 3' followed by falls and rises of several feet within a few minutes. Severe thunderstorms were over west Lake Ontario at the time.

8 2004: An alia (fishing boat) was upended by a huge wave near the beach at Fogagogo (American Samoa), causing the engine to fall off. The alia's 4 fishermen were not seriously hurt. High swells were caused by strong high pressure far to the south.

9 2002: Vehicles are not always safe refuges from hailstorms. Wind-driven hail (to 1.5" diameter) broke windows of a pickup near Hertick, SD; a man sitting inside the truck was injured by high-velocity hail. Hail in the area accumulated to 6" deep, drifts to 2' deep.

10 1973 (10–12): Forest fires and avalanches can lead to potential landslide problems when heavy rains fall on the denuded slopes. Heavy rains led to mudslides, which covered 2 miles of Thane Road (Juneau, AK, area) at the site of a snowslide on January 22nd this year.

11 2004: If your car is stalled in high water, look for downed power lines before exiting the vehicle. At College Point, NY (NYC metro area), 2 people were electrocuted by a fallen power line when they left the car and stepped into water several feet deep.

12 1972: In Mississippi's Sunflower County (near Ruleville), lightning struck and killed 1 of 2 boys walking down a turnrow. The county coroner said the 2nd boy was not seriously hurt, "the lightning just knocked him down and he jumped up and ran."

13 2002: A heatburst hit San Angelo, TX, just before midnight. Downdrafts from dying thunderstorms warmed the temperature from 75°F at 11:35 P.M. CST to 94°F at 12:05 A.M.; humidity fell from 62 to 19%; wind gusts to 40 mph. By 12:30 A.M., readings back to 73°F and 66%.

14 1946: Southern U.S. has no concern with freezes this time of year; not so with southern AK. Today's 32°F low at Anchorage became its earliest autumn freeze on record. Its all-time record high: 86°F on 6/25/1953.

15 1958: Fortunately, most tornadoes are relatively weak. A cooperative weather observer's pickup truck was lifted 20' and dropped upright by a tornado in a forested part of Easton, CT. He noted a swirling mass of limbs/debris some 100' above ground.

16 1972: A neutercane (term retired in 1973; subtropical cyclone is now used) deepened offshore Crescent City, CA, and made landfall near Brookings, OR. Winds 70 mph sank at least 10 boats, damaged 100. 13 drowned (4 in CA waters, 9 in OR waters).

17 1969: Hurricane Camille (only 1 other modern-day storm a Category 5 at U.S. landfall) moved inland west of Pass Christian, MS, with winds to 200 mph and a 24.6' storm surge. Total destruction in some coastal areas near eye. 144 plus Gulf Coast deaths.

18 1972: At Kapoho Vacationland (on the Puna Coast, Island of Hawaii), waves to 30' high (generated by Tropical Storm Diana) swept 4 homes from their foundations. Waves knocked down a couple and swept them into bushes, but did not pull them out to sea.

19 1972: Hurricane Celeste passed 25 miles northeast of Johnson Island, an atoll 1 square mile in size with highest point <20' above sea level; winds more than 100 mph. The civilian and military population of 500 was evacuated (18th) due to the danger of stored toxic gases escaping.

20 1999: A waterspout from the Atlantic Ocean moved onshore as an F2 tornado at Beach Haven, NJ; $4.2 million damage. The Sea Spray Motel was severely damaged (sprayed with more than sea water); its 150 vacationers were displaced.

21 2003: August is summer in the northern Hemisphere but winter in the southern Hemisphere. This was likely the coldest August night on record for South Africa as new monthly records were established throughout the country. Johannesburg's 19°F set its new record.

22 2002: Thunderstorm winds to 80+ mph tore a l00'-long blimp from its berth at Timmerman Field (Milwaukee, WI); it flew 6 blocks before impacting 4 homes. Windblown rain came through Miller Field's retractable roof—saturating the baseball field and some spectators.

23 1999: Video games provide excitement, but this may have crossed the line. At Sanford, ME, a man playing a video game in the American Legion building was injured when a lightning strike arced through all the video machines. He suffered chest pains and hearing problems.

24 2005: Tropical Storm Katrina formed in the central Bahamas. Katrina would become the costliest weather disaster in U.S. history: damage estimates more than $100 billion. Flooding of New Orleans, LA, would displace more than 250,000 people, more than displaced in the 1930s' Dust Bowl.

25 2005: Hurricane Katrina made its 1st U.S. landfall on FL's east coast between Hallandale Beach and North Miami Beach; sustained winds of 80 mph, gusts of 90+ mph. Katrina killed 14 as it crossed extreme southern FL, including several killed by falling trees and tree branches.

26 1999: 2 to 6" rain pounded the NYC metro area; mass transit crippled during the morning rush. Water 3 to 5' deep at some subway stations; part of the N-bound platform of the 6 line at 28th Street washed away. Some parkways closed; trains with thousands of riders stranded.

27 1999: Moisture from remnants of Hurricane Bret ignited thunderstorms that raked west side of Phoenix, AZ. Microburst wind/rain led to the evacuation of several thousand people from the Desert Sky Mall; parts of the roof collapsed 10 minutes after the evacuations.

28 2005: Hurricane Katrina's peak intensity of 175 mph sustained winds occurred about 225 miles south/southeast of the mouth of the Mississippi River. The 4 P.M. National Hurricane Center bulletin warned of Gulf Coast storm surges to 28' and "Some levees in the Greater New Orleans Area could be overtopped."

29 2005: Hurricane Katrina made its 2nd U.S. landfall just south of Buras, LA, as a Category 3 storm; sustained winds 125+ mph. New Orleans Lake Front Airport reported sustained winds of 69 mph, gusts to 86 mph. The LA/MS death toll (mostly surge/flooding) was more than 1,300.

30 1979 (8/25–9/7): Hurricane David's lowest pressure (924 mb) and highest sustained winds (173 mph) were south of Puerto Rico today. 60,000 of the island of Dominica's 80,000 residents were left homeless by David's direct hit (29th); 56 deaths there.

31 1954: Winds to 86 mph from Hurricane Carol felled the steeple atop Boston's (MA) Old North Church (in which lanterns were hung to tell Paul Revere the British were coming). The steeple was built in 1806 to replace the original steeple toppled by an 1804 hurricane.

SEPTEMBER

1 1979: In the Houston (TX) Ship Channel the oil tanker Chevron Hawaii was hit by lightning from storms of Tropical Storm Elana; 3 killed, 4 injured in ensuing explosion/fire. Oily rain covered lawns/houses downwind (from smoke combining with raindrops).

2 1974: At about 1:30 A.M., a weak tornado skipped across parts of the Bronx and Mount Vernon, NY. The path width varied between 50 and 200 yd.; its length was 2 miles. It caused minor damage and no injuries.

3 1997: Thanks to 4.5" rain, a 12' high wall of water washed over AZ-14 and the Red Rock-Randsburg Road near Cantil, AZ; 4 cars were swept away, but no deaths; 100 motorists stranded. Much flooding in Red Rock Canyon State Park. $5 million damage; 4 injuries.

4 2004: Astronomical summer can be anything but "warm" in the high Rockies. A hiker froze to death on the summit of CO's Longs Peak, as a not rare snowstorm with high winds swept the area. A hooded sweatshirt, jeans, and tennis shoes were his only cold weather gear.

5 1933: A hurricane made landfall just north of Brownsville, TX. A wind gage blew away at 106 mph; estimates to 130 mph. Drifted citrus blocked roads; reports of houses floated 10 miles by surge. Some farmland within 10 miles of bay still unproductive due to salty surge.

6 1997: Lightning can strike any time of the day. A Saturday morning football game was marred at 8:35 A.M. in Fruita, CO, when lightning hit 2 power poles at a middle school. Several cheerleaders were knocked to the ground by the shock wave.

7 2000: Lightning hit near a basketball hoop at Oceanside (CA) High School as 200 students met for PE class. Two knocked unconscious said it was as if hit by a rock or needles through their skin and hair pulled. 100 students "jolted"; hair on arms and heads stood on end; 3 injuries.

8 1900 (8–9): America's worst natural disaster (number of deaths) occurred when a hurricane with 120 mph winds and a 20' storm surge killed 6,000 in Galveston, TX. 1,200 died elsewhere. Following the storm, the surf was 300' inland from the former waterline.

9 2004: Close call for 17 campers near Conneaut Creek in OH's Ashtabula Counry. Rain (nearly 5") caused a rapid rise in water and put a campground under 5' of water shortly after midnight. The campers barely escaped the water's rise and had to be rescued by boat.

10 2004: A wall of water and mud to 10' high raced down CA's Borrego Palm Canyon and into Borrego Springs; a campground (fortuitously empty) was obliterated. 70 to 90 homes were damaged in Borrego Springs by mud to 2' deep. $1 million damage.

11 1998: 2" rain fell on parts of UT's Zion National Park in 24 hr. Flow in the Virgin River at the park headquarters went from 280 to 4,200 cfs in 4 hr. The raging water created a sinkhole 30' long by 15' wide by 20' deep in the Zion Canyon Scenic Drive.

12 1976 (11–12): Japan's 24-hour rainfall record of 44.80" at Hiso (Tokushima Prefecture) was caused by Typhoon Fran. Fran was also responsible for 167 deaths in Japan.

13 2000: A hurricane hazard sometimes not in mind. Hurricane Florence, far beyond the horizon, nonetheless generated large swells that caused rip currents along NC beaches. Two people drowned an hour apart at Kure Beach, and another at Carolina Beach.

14 1974: A flash flood roared down NV's El Dorado Canyon into Nelson's Landing. A trailer park (35 house trailers), bar, restaurant, 50 vehicles, and part of a boat landing were washed into Lake Mojave. Nine deaths.

15 2004 (13–16): Hurricane Ivan affected coastal AL/W FL Panhandle; landfall occurred near Gulf Shores, AL, early on 16th. A buoy just south of the AL coastal waters reported an incredible peak wave height of 52' today before breaking loose of its mooring.

16 2004: Hurricane Ivan made landfall near Gulf Shores, AL, early morning. Some deaths after the storm's end: a 7-year-old boy killed by a falling limb while he watched a tree being removed; an 83-year-old man fell off a damaged roof he was repairing.

17 2004 (16–17): 90,000 apple trees were blown down by remnants of Ivan in NC's Henderson County. On 16th, a debris flow of water/trees/boulders/mud from the top of Fishhawk Mt. destroyed/damaged 20 to 30 homes/mobile homes in Peeks Creek Valley; 4 died near Macon.

18 2004: WY's Wind River Basin lived up to at least part of its name. A wild fire just south of Riverton, spread by 30 to 35 mph southwestern winds, burned 300 acres, 33 vehicles on a used car lot, 4 campers, and a mobile home used for storage; 9 injured; $1 million damage.

19 2000: Sure looked like a western wildfire, with trees and houses burning and some flames shooting skyward more than 50' high. But instead, the wildfire was just north of Oklahoma City, OK (from 9 miles south to 3 miles south of Guthrie); 35 homes destroyed.

20 1962: Near Fruitland, NM, lightning hit the tin chimney of a hogan during a native ceremony; 2 killed, 2 injured. Yesterday, at Missoula, MT, "dry" lightning (meaning little, if any, rain was produced by the storm) burned a grain elevator; 50,000 bushels of wheat ruined.

21 1962: Amazing what lightning can do when it hits a person, yet not cause death. At McCammon, ID, the zipper fastener on a schoolboy's jacket and trousers were fused by a lightning strike. He did suffer severe burns.

22 1995 (21–22): Some things are out of a farmer's control. Record early cold and freezing temperatures covered most of IA overnight; as cold as 24°F this morning in Emmetsburg and Sibley. 30 million bushels of soybeans lost, along with some corn. $204 million crop loss.

23 1997 (23–25): Surf as high as 20' pounded HI's north beaches; the surf was generated by what had been Typhoon David in the north Pacific. 30+ surfers, bodyboarders, and swimmers were rescued along Oahu's Sunset Beach and Waimea Bay.

24 2001: 10 trailers on the campus of the U of MD were destroyed by an F3 tornado. One occupant was thrown into a nearby dumpster; others "dug their hands into the carpet" and held on for dear life as the tornado pulled their feet up in the air.

25 1997: Add El Niño in the east Pacific, large swells from a distant storm in the Gulf of AK, effects of Hurricane Nora, and high tide. The result: 5 to 8" waves moving onshore and flooding a 14 block stretch of Seal Beach, CA. Quick response by the city crews limited damage.

26 2004: Thunderstorm rains closed part of TX-118 with water to 6' deep between Fort Davis and Fort Davis State Park. Two campers were trapped in Short Canyon for almost 24 hour. They were rescued by helicopter on 27th at a deserted ranch house near Wild Rose Pass.

27 1962 (27–28): Gale-force winds/heavy rains pounded ME. Tides to 4' above normal; falling trees downed many power and phone lines. At Fort Fairfax, a close call: Strong winds blew a boy and the canopy on which he was lying from a moving truck; no serious injuries.

28 1962: Seattle, WA's, 1st known tornado (a waterspout over Lake Washington) caused mainly minor damage in small areas of the northeast part of the city. However, a house roof was severely damaged in the View Ridge area, as was a hangar roof at Sand Point Naval Air Base.

29 2004: Although Hurricanes Charley (in August) and Frances/Ivan/Leanne (in September) had exited FL, their effects were long lasting. According to the Insurance Information Institute, 1 of every 5 houses in FL received damage from 1 or more of these storms.

30 1875: What whether observers "observe" has changed with time. Among the remarks noted by the Lynchburg, VA, observer, was today's migration of hundreds of squirrels across the James River.

OCTOBER

1 1986: All it took was an estimated 5 seconds for thunderstorm winds of 75 mph to lift two aircraft at TN's Tullahoma Airport, and then dash them into the ground upside down. One was totally destroyed. Two hangars were also damaged.

2 1858: Wind, rain, and damage accounts show the only known tropical system to hit southern CA as a hurricane produced Category 1 winds at San Diego and tropical storm force winds at Long Beach. A similar storm in 2007 might produce $200 million damage in the same area.

3 1964: Hurricane Hilda produced an F4 tornado, which tracked west/northwest 2 miles through LA's Lafourche Parish. 22 people killed in Larose; debris blown 16 miles to the west. 38 people killed by Hilda; 22 by this tornado.

4 1777: Dense fog cloaked the battle of Germantown (PA). Smoke from the fight combined with fog to make it "almost as dark as night." Due in part to the poor visibility, Americans at times shot each other, helping the British eventually win the battle.

5 2004: Baseball-sized hail pounded Socorro, NM, for 5 to 10 minutes. The result was $40 million damage in Socorro County, of which $15 million occurred on the New Mexico Tech campus. Nearly every university building was damaged, and the university vehicle fleet was almost a total loss.

6 1981: A Fokker F-28 aircraft flew into thunderstorms near Moerdijk, Holland. The detachment of its starboard wing led to a crash with the loss of all occupants. The plane had likely flown into a funnel just after the tornado lifted from the ground.

7 1997: Did he take his job home with him? At Mesa, AZ, hail accumulated to 6" deep. An employee of the National Weather Service reported roof tiles and a glass window were broken on his house.

8 1972: Between Guam and Rota in the Pacific Ocean, Typhoon Marie killed 2 men. Three men assumed dead were found 47 days later, 1,310 miles southwest of Guam by a Japanese fishing boat. In good health, they survived by eating fish and wringing rainwater from their clothes.

9 2005 (7–9): Remains of Tropical Storm Tammy helped dump 7 to 12+" rain on parts of NC's Robeson/Brunswick/New Hanover Counties. Much home and business flooding; many roads impassable; $2.5 million damage. Basement of Cape Fear Museum (46,000 artifacts stored) completely flooded.

10 1990: A 50' section of Kendall Lake Dam burst just upstream of Camden, SC. Motorists trying to push cars through waist-deep water were swept 50 to 100 yards downstream by the additional torrent; some held onto trees until rescued; 4 deaths in 1 stalled car.

11 2005: Ex-Hurricane Vince weakened to a tropical depression at 0000 UTC 155 miles west and southwest of Faro, Portugal. It passed just south of Faro and made landfall near Huelva, Spain. It is the only known tropical cyclone to reach the Iberian Peninsula. No casualties/serious damage.

12 1886: Landfalling hurricane between Sabine Pass, TX, and Johnson's Bayou, LA. Tidal waves said to be as high as 2-story buildings; surge extended 20 miles inland. 150 killed; survivors clung to trees and floated on mattresses. Two of 100 homes in Sabine Pass repairable.

13 1991 (13–14): A deep low-moving storm on the east side of the Kenai Peninsula helped give Anchorage, AK, its October record 24-hour snow of 11.6" (old record: 10" on 10/10/1940).

14 1966: Late-season severe thunderstorms brought up to softball-sized hail to parts of Brown, Dodge, Freeborn, Pipestone, Ramsey, Rock, and Steele counties in MN. A hailstone reportedly 16" circumference (MN record if true) smashed a truck windshield near Claremont.

15 1996: Waterspouts are not always confined to warm, tropical waters. Convective showers, which developed in a cold, unstable air mass, produced a waterspout over the Pacific Ocean off Seaside, OR. No damage occurred.

16 1984: An F1 tornado (25 yards wide with a 1-mile-long path) struck Lawler, IA, significantly damaging 4 homes. One resident said, "There was so much suction . . . we couldn't get the door to the basement open."

17 2005: Tropical Storm Wilma formed 0600 UTC 170 miles southeast of Grand Cayman. By 19th, Wilma explosively deepened to become the most powerful Atlantic hurricane of record with a pressure of 882 mb (26.05") and sustained winds of 185 mph 340 miles southeast of Cozumel, Mexico.

18 2004: A good day not to have been fishing. An F1 tornado near Higdon, MO, picked up a fishing boat from a pond, carried it 100 yards and dropped it in some trees. The tornado also wrapped a trampoline around a tree and left up to 2" of leaves inside a damaged home.

19 1992: Driving blind/driving too fast for the conditions? Dense fog (visibility often less than ¼ mile) led to a 30-car and 1 18-wheeler chain reaction accident on CA's I-15 near Rancho Cucamonga; 24 injuries. Road closed for more than 10 hours.

20 1991 (20–22): Strong northeast Diablo winds fanned a small brush fire into an inferno that destroyed more than 3,300 homes in the hills of Oakland and Berkeley, CA; 25 died and 150 were injured. Property damage exceeded $1 billion.

21 2003 (21–29): The Roblar 2 fire was started on the Camp Pendleton Marine Corps Base (CA) by live ammunition training; 8,592 acres burned before containment. Firefighters (more than 1,300) had to watch for unusual hazards—unexploded ammunition.

22 1996: Kansas City (MO/KS) whitened by a general 6-9" snow, its earliest and heaviest of record. The 6.1" observed at the International Airport became the official October single snowfall record for the city. Much tree and power line damage (175,000 people without power).

23 1997: An F1 tornado rolled a mobile home near Jasper, TX, 100 yards up a hill before it hit a tree. Two people inside were not seriously hurt because they had taken shelter in the bathtub shortly before the tornado hit the trailer.

24 2001: Fortuitous timing. An F1 tornado destroyed 4 classrooms and heavily damaged 8 others at the Davisburg (MI) Elementary School. The tornado hit 8:10 P.M. EDT. The only other significant damage caused by the storm was some uprooted trees.

25 1997: Another good reason for having a smoke detector in your home! About 9:30 P.M., near Hartselle, AL, lightning struck a switchbox, traveled into the house, and set the house on fire. A man inside the house died of smoke inhalation.

26 1992: 2 men working at a rock quarry near Cookson, OK, were attaching caps to 24 sticks of dynamite when a bolt of lightning struck about ¼ mi. away. The resultant ground currents ignited the dynamite; the men died.

27 1959: The ship Mary Barbara reported a 958 mb pressure and estimated winds to 161 mph just outside Manzanillo (Mexico) Harbor. This "Manzanillo" hurricane was one of Mexico's worst; 1,000 deaths in and around Mammillo, Minatitlan, and surrounding areas.

28 1999 (28–29): Northwest swells caused high surf on CA's central/southern beaches. 40' high waves broke through the seawall at Capitola; some streets and businesses flooded. A 25' wave swept a honeymooning couple to sea (29th) from Lover's Point Beach (Pacific Grove); 1 killed.

29 1956: With a prolonged and deafening noise, an F3 tornado destroyed and damaged buildings west of Dorrance, KS; most were rebuilt after a 1951 storm. Near Hunter, a farmhouse was "utterly demolished." A mother and baby were blown 60' into a field, but had only bruises and scratches.

30 1993: Snow began very late (29th) over southwest IN and spread into southwest OH and west and north KY. Evansville, IN, whitened by a 24-hour October record 4.6". Today's 5.9" at the Cincinnati/N KY Airport set its October record and gave Cincinnati its first "white" Halloween.

31 1965: A major windstorm plagued the west half of NY and adjacent Great Lakes; wind gusts were to 80 mph near the Great Lakes and 60 mph inland. The St. Lawrence Seaway was closed for 11 hours, delaying ferry traffic between the U.S. and Canada.

NOVEMBER

1 2002 (10/29–11/4): Frigid temperatures followed the passage of an Arctic cold front across WA on October 29th. About 3.9 million bushels of apples were killed by the cold in Yakima Valley (lows to 4°F); $65 million loss to farmers.

2 1967: The Last Chance Motel in Emery, UT, was destroyed by a cone-shaped F2 tornado; furniture and bedding were thrown hundreds of yards. The tornado occurred at the unusual time (for a tornado) of 8:30 A.M. No injuries from this event.

3 2004: Winds to 65 mph blew down a number of trees in OR's Crater Lake National Park. Falling trees damaged the community center, the Good-by picnic area comfort station, a house in Sleepy Hollow, and part of the Science and Learning Center.

4 1988: A tornado touched down briefly near a school in McComb, MS. The tornado missed a filled football stadium by only 300 yards.

5 2002: Wind gusts to 70 mph at high tide led to $1 million damage to the Larsen Bay dock area on west side of AK's Kodiak Island. A tidal surge of 1 to 2' on top of the high tide caused waves that "smashed" the dock and pier and mess hall/bunkhouse floors from below.

6 1953 (6–7): Earliest, heaviest snow of record for much of the East Coast. Some records: 11.9" at Wilmington, DE; 6.5" at Washington, DC; 3.7" at NYC, NY. Greatest storm amount was 27.5" at Middleburg, PA. Many roads in central eastern PA closed by 3 to 4' drifts.

7 1998: Snowboarding and skiing can be risky fun. Five snowboarders were swept down a rocky slope of UT's Little Cottonwood Canyon of the Wasatch Mountains; 1 killed, 1 injured. Nearby resorts had not opened and avalanche controls had yet to be exercised.

8 1996: Tornadoes are relatively rare in PA; November tornadoes even more so. Nonetheless, an F1 tornado, with numerous, brief touchdowns, tracked 10 miles from near Kunkletown to near Syndersville. A house roof was blown 100 yards. Fortunately, no injuries.

9 2001: A wildfire spontaneously combusted in a 4-acre stump pit (50' deep) near Marshalls Creek, PA. Windy, dry weather combined with fallen leaves to spread the fire. Dense smoke led to potential respiratory problems for nearby residents. Fire put out on 12th.

10 2002: F0/F1 tornadoes are "weak" tornadoes; they are responsible for only 5% of tornado deaths, although they comprise 70% of all tornadoes. As an F1 tornado traveled through Crawford, MS, it killed a man as he attempted to warn neighbors of the approaching storm.

11 1987 (11–12): Heavy snow (to 17") whitened Washington, DC, and adjacent areas of VA/MD. Some motorists spent the night in their vehicles on I-295 or on the Woodrow Wilson Bridge; some students spent the night in their Prince George's College (MD) school.

12 1956: The icebreaker USS *Glacier* saw what may have been the world's largest iceberg. Seen about 150 miles west of Antarctica's Scott Island, the iceberg was about 60 miles wide by 208 miles long, or roughly the size of MD. (Date approximate.)

13 1992: The Peace Bridge between Buffalo, NY, and Fort Erie, ON (Canada), shut 7 hours when high winds toppled 2 semi-trailer trucks crossing the span over the Niagara River. The same winds caused Lake Erie levels to rise 3' in 1 hour on its eastern side; much flooding.

14 2001 (13–14): Smoke from a brush fire that started on the 11th in NJ's Pennsville Township became trapped under an inversion. Thick smoke caused schools to close, and closed several major roads in the A.M.

15 2001: An F0 tornado traveled 2 miles just northeast of Austin, TX; trees destroyed/carports and mobile homes damaged. A school in its potential path was having a tornado drill. With students in hallways, no further action needed. The tornado lifted before hitting the school.

16 2001: After several days of rain, dense fog developed in low-lying areas of south-central/southeast WI. The driver of a vehicle was killed when it struck a horse standing on WI-78 near Mazomanie; visibility was 10'. Sixteen children hurt when a Monticello school bus was hit by a truck.

17 2002 (16–17): A major ice storm glazed northern CT with up to ¾" ice; $2.5 million damage, mainly from falling trees. Towns such as Granby, Simsbury, and Canton without power for up to 5 days. 50 mph winds on 18th blew down trees and power lines still coated in ice.

18 1995: Winds to at least 89 mph caused much damage in central MT. Vehicles were blown off roads near Moore, Sunburst, Hays, and Brady. Roofs were damaged and many signs blown over.

19 1986: Early season snow records for parts of southestern New England: 11" at Blue Hill Observatory (Milton, MA) beat the previous record of 8"; 4.2" at Providence, RI, set its early record. Even 1" fell on Cape Cod, MA. Storm on November 11 to 12, 1987, will establish new early records.

20 2001: Today's 31°F low is the latest 1st fall freeze on record for Albuquerque, NM (airport records began in 1931). An astonishing 80 of 91 fall lows (meteorological fall is Sept. through Nov.) were normal for Albuquerque.

21 2004: An avalanche hit 3 ice-climbers in First Gully near Eureka, CO. The lead climber fell 200' and was buried under 6" of debris; his back was broken. A second climber was partially buried up to his waist in a sitting position. All survived.

22 1993: Reduced visibility due to blowing snow, along with snow and ice, caused a 90-car pileup on I-90 in Spokane, WA, just before 11 A.M. 45 injuries; $500,000 damage. I-90 was closed for up to 5 hours.

23 2001: An F2 tornado passed from near Hunt to Salus in AR; 1 person killed in a mobile home. North of Hunt, 2 people were injured in 1 of 6 chicken houses destroyed at the location; 3 of the chicken houses destroyed contained 120,000 chickens; most perished.

24 1940 (23–25): Said to be worst ice storm in U.S. history through 1940, freezing rain and drizzle fell on north TX Panhandle. Ice on power lines to 6" circumference; weight of 13 lb/linear ft downed 1,000s of poles/trees. All power lost in Amarillo/Canyon for 3 days.

25 1985: Tornadoes are much rarer in South America than in North America. Today's tornado in Argentina's east Buenos Aires province tracked 25 miles; its width was up to 325 yards wide. It produced a lot of damage at Dolores, where 1 person was killed.

26 1995 (26–27): Heavy snow closed I-70 between Eisenhower Tunnel and Vail Pass, CO for several hours, stranding hundreds of travelers going home from ski resorts at the end of Thanksgiving weekend. The American Red Cross opened many temporary shelters.

27 1997: A windy day in NYC, NY, for the Macy's Thanksgiving Day Parade. The strong winds (50 mph gusts) led to balloon handlers losing control of the Cat in the Hat balloon. It hit the top of a light pole, which fell on 4 onlookers; all hurt, 1 seriously.

28 1912 (27–28): Snowfall is a rare event in FL. The record earliest snowfall for FL (and its only November snow) fell during the night across parts of the interior counties in the far northwest (Madison to Gadsen) bordering GA. Up to ½" was reported at Mt. Pleasant.

29 2001: Flash flooding widespread across central/north-central TN; a number of roads were closed. The only wooden bridge left in Stewart County was washed away. Many schools closed because of flooded roads.

30 1997: Based on medical reports, at Milwaukee, WI, an 88-year-old woman found dead on her driveway died of hypothermia. Low temperatures were in the mid-high 30s. It is speculated she fell down and was unable to get up.

DECEMBER

1 2004: Winds to 50 mph caused trees and wires to topple in Bath, NY. A 10' high facade of a building was blown down, falling on a woman 15' below. Her rescuer was injured by continuing debris fall.

2 1982: Today became Chicago's warmest winter day on record with a high of 71°F and a low of 58°F.

3 1973 (3–4): An ice storm in extreme southeast New England toppled 1,000 poles, uprooted trees, and cut power to 25 towns (worst damage southeast of a line from Superior to Plattsmouth). Some rural areas without power for 4 to 6 days.

4 1973: In the Braman area of IN, a tornado "only 5–10' wide and 50' high" damaged a garage, overturned a boat and trailer, and threw a pontoon boat into trees. A truck on US-6 was lifted from the roadway and slammed into a tree; 1 injury.

5 1975: An F2 tornado hit the gym at Eastern Oklahoma State College in Wilburton, OK; 200 people were in the gym, but only 3 were hurt (by flying glass). Good luck continued as the tornado hit the west wing of a nursing home, but injured no occupants.

6 1997: Biggest rains in at least 70 years totaled 4 to 8" in CA's central Orange County; 10" at Mission Viejo. Hillsides collapsed; many mud/debris flows; much flooding. Areas such as Newport Bay's Rhine Channel so clogged they looked like landfills. $17.7 million damage.

7 1950: All American Airlines at Connellsville, PA, observed winds of 60 mph with gusts to 75 mph. At least 8 vehicles damaged by falling buildings. A boy was blown into a creek, but rescued.

8 1973: Near Kulani Honor Camp (off Stainback Highway on island of Hawaii), strong winds toppled an ohia tree on a hunter and pinned him; his companion was unable to lift the tree and went for help. The pinned hunter was killed by more falling trees before help arrived.

9 2003: Although a "fish" storm (it never threatened land), a subtropical storm became Tropical Storm Peter 700 mi. west/northwest of the Cape Verde Islands. This is the first time since 1887 that 2 tropical storms formed in the Atlantic Basin in December.

10 1995: In frigid pans of the U.S., motorists are advised to have cold weather survival gear in their vehicles. 30 mi. southeast of Ontonagon, MI, a man was found frozen to death in a snowbank near his disabled vehicle. Temperature was as low as –15°F, windchill to –60°F.

11 2002: An unusual event in NJ's Lafayette Township: a satellite dish repairman had to be rescued from the roof of a building after it iced over while he was working on the dish. Rain turned to ice on contact with the cold roof surface.

12 1995 (11–12): A Pacific storm raked OR; gusts: 119 mph (Florence), 112 mph (Cannon Beach), 101 mph (Astoria), 74 mph (Portland, downtown windows blown out). Falling trees killed 2; a livestock gate caught by the wind killed a woman opening it. $25 million damage.

13 1987 (12–14): Heavy snow and high winds raged across most of NM; much of the interstate system closed. Snow totals to 3' in higher elevations. Unofficial wind gust at Sandia Peak Tramway northeast of Albuquerque of 124 mph this P.M.; Coronado Airport wind gust of 107 mph.

14 2005: At 5:15 A.M. a dam failed; 1.3 billion gallons of water raced through MO's Johnson's Shut-Ins State Park; most facilities destroyed or damaged. Park empty; in spring and summer often several thousand daily users. The annual dam break exercise is scheduled on this day.

15 2005: A major ice storm glazed much of the piedmont and adjacent mountain areas of GA, NC, and SC with up to ¾" accumulations. Many trees and power lines downed. 683,000 customers lost power. Fortunately, temperatures were not bitterly cold; mostly close to freezing.

16 1997 (15–16): Although Super Typhoon Paka left major damage to 60% of homes (1,160 single family homes destroyed) on Guam (wind gusts to 175 mph before instruments failed), 0 deaths and only 2 hurt. $500 million damage.

17 1973 (16–17): The worst ice storm in CTs history paralyzed central/western sections of state. Statewide tree damage was worse than damage caused by infamous 1938 hurricane. More than 250,000 homes were without power and heat for up to a week.

18 1971: Hail is unusual on the tropical islands of HI. Nonetheless, unusually large hail (for HI) the size of "marbles" and "half an inch in diameter and two inches long" was reported at Kealia (Kauai).

19 1978 (18–19): Unseasonably heavy rains of 1 to 3" on top of saturated ground and mountain snowpack led to record crests of NM's San Francisco, Gila, and Mimbres Rivers. The Gila, normally 5 to 6' wide at Redrock, was "nearly one mile wide" at crest this morning.

20 1977: Winds to 80 mph snapped a power pole on CA's Vandenberg Air Force Base; ensuing brush fire consumed 10,000 acres before brought under control. 450 plus firefighters; 30 bulldozers fought the fire. The base commander and 2 top base fire officials killed by the fire.

21 2002 (21–24): An unusual combination of rain and hail (hail is rare at low latitudes) caused flooding and landslides on Vanuatu's Tanna Island (southwestern Pacific); some bridges washed away. Many food gardens destroyed; 3,000 plus people affected.

22 1998 (21–25): Citrus/vegetables in CA's San Joaquin Valley seriously hurt by 5 nights of freezing temperatures. Today's 34°F high in Bakersfield its coldest-ever high, breaking the 35°F reading of 12/11/1932. Contrast this to today's record 32°F high in Battow, AK.

23 1973 (23–24): A blizzard played havoc with pre-Christmas travel in east CO; 11.8" at Denver's Stapleton Airport, where 10,000 people were stranded for 24 hours. Many roads blocked by drifts to 10' deep.

24 2001 (24–28): Almost no snow this season at Buffalo, NY, until this 5-day storm. 25.2' snow in 24 hours. (24–25): 4th largest on record; 35.4" in 24 hours (27–28): 2nd largest on record. 81.6" new storm total record; depth of 44" (28) record depth.

25 2001: Part of IN-257 in Pike County was barricaded due to flooding. At least 4 vehicles whose drivers ignored the barricades were swept away. One death likely late evening; vehicle found the 26th.

26 1996: An ice storm knocked out power to 269,000 people in WA's south Puget Sound region; Tacoma Narrows Bridge closed. Up to 10" snow in Seattle/Everett. 400+ flights canceled at Seatac Airport. Almost all hotels from Seatac to Tacoma full with stranded people.

27 2005: This December was OK's 4th driest (back to 1895). Wildfires burned 175,000 acres; 150 homes and businesses destroyed or damaged; $10+ million damage. Today's fires in Hughes County burned 10,000 acres. A 68-year-old male died while firefighting.

28 1978: Severe thunderstorms produced strong winds and golf-ball-sized hail in FL's northern Dade County. A 25-year-old jockey was struck and killed by a lightning bolt at Calder Race Course. He had just finished a race and was running for cover.

29 1973 (29–30): An F3 tornado ended its path at Fort Rucker, AL, causing only minor damage. The next day an F3 tornado caused $1 million damage to government buildings and a residential area in Fort Rucker, and injured 23 people.

30 2005: Tropical Storm Zeta became the 27th named storm of the 2005 Atlantic season, shattering the old record of 21 (1933). Although computer models kept advertising Zeta's demise, the storm managed to survive in the open waters of the central Atlantic through 1/6/06.

31 1968: Not a good way to end the year. Heat in an icehouse on MN's Lake Beebe was lost in cold, windy weather; wind chill down to –53°F. Three boys ice fishing suffered severe frostbite; 2 of them were in critical condition for a time.

GLOSSARY OF METEOROLOGICAL TERMS

ACCAS AltoCumulus CAStellanus; mid-level clouds (bases generally 8 to 15 thousand feet), of which at least a fraction of their upper parts show cumulus-type development. These clouds often are taller than they are wide, giving them a turret-shaped appearance. ACCAS clouds are a sign of instability aloft, and may precede the rapid development of thunderstorms.

Advection Transport of an atmospheric property by the wind.

Air-mass Thunderstorm Generally, a thunderstorm not associated with a front or other type of synoptic-scale forcing mechanism. Air-mass thunderstorms typically are associated with warm, humid air in the summer months; they develop during the afternoon in response to insolation, and dissipate rather quickly after sunset. They generally are less likely to be severe than other types of thunderstorms, but they still are capable of producing downbursts, brief heavy rain, and (in extreme cases) hail over ¾ inch in diameter.

Aurora It is created by the radiant energy emission from the sun and its interaction with the earth's upper atmosphere over the middle and high latitudes. It is seen as a bright display of constantly changing light near the magnetic poles of each hemisphere. In the Northern Hemisphere, it is known as the aurora borealis or northern lights, and in the Southern Hemisphere, this phenomena is called the aurora australis.

Barotropic System A weather system in which temperature and pressure surfaces are coincident, i.e., temperature is uniform (no temperature gradient) on a constant pressure surface. Barotropic systems are characterized by a lack of wind shear, and thus are generally unfavorable areas for severe thunderstorm development.

Blizzard A blizzard means that the following conditions are expected to prevail for a period of 3 hours or longer: Sustained wind or frequent gusts to 35 miles an hour or greater; and considerable falling and/or blowing snow (i.e., reducing visibility frequently to less than ¼ mile).

Cap (or Capping Inversion) A layer of relatively warm air aloft (usually several thousand feet above the ground) which suppresses or delays the development of thunderstorms. Air parcels rising into this layer become cooler than the surrounding air, which inhibits their ability to rise further. As such, the cap often prevents or delays thunderstorm development even in the presence of extreme instability. However if the cap is removed or weakened, then explosive thunderstorm development can occur.

The cap is an important ingredient in most severe thunderstorm episodes, as it serves to separate warm, moist air below and cooler, drier air above. With the cap in place, air below it can continue to warm and/or moisten, thus increasing the amount of potential instability. Or, air above it can cool, which also increases potential instability. But without a cap, either process (warming/moistening at low levels or cooling aloft) results in a faster release of available instability—often before instability levels become large enough to support severe weather development.

Cb Cumulonimbus cloud, characterized by strong vertical development in the form of mountains or huge towers topped at least partially by a smooth, flat, often fibrous anvil. Also known colloquially as a "thunderhead."

Cell Convection in the form of a single updraft, downdraft, or updraft/downdraft couplet, typically seen as a vertical dome or tower as in a towering cumulus cloud. A typical thunderstorm consists of several cells. The term "cell" also is used to describe the radar echo returned by an individual shower or thunderstorm. Such usage, although common, is technically incorrect.

Celsius Temperature Scale A temperature scale where water at sea level has a freezing point of 0°C (Celsius) and a boiling point of +100°C. More commonly used in areas that observe the metric system of measurement. Created by Anders Celsius in 1742. In 1948, the Ninth General Conference on Weights and Measures replaced "degree centigrade" with "degree Celsius." Related term: Centigrade

Cirrus One of the three basic cloud forms (the others are cumulus and stratus). It is also one of the three high cloud types. Cirrus are thin, wispy clouds composed of ice crystals and often appear as veil patches or strands. In the mid-latitudes, cloud bases are usually found between 20,000 to 30,000 feet, and it is the highest cloud that forms in the sky, except for the tops, or anvils, of cumulonimbus, which occasionally build to excessive heights.

Climatology The study of climate. It includes climatic data, the analysis of the causes of the differences in climate, and the application of climatic data to the solution of specific design or operational problems.

Cold-air Funnel A funnel cloud or (rarely) a small, relatively weak tornado that can develop from a small shower or thunderstorm when the air aloft is unusually cold (hence the name). They are much less violent than other types of tornadoes.

Confluence A pattern of wind flow in which air flows inward toward an axis oriented parallel to the general direction of flow. It is the opposite of difluence. Confluence is not the same as convergence. Winds often accelerate as they enter a confluent zone, resulting in speed divergence which offsets the (apparent) converging effect of the confluent flow.

Convection Generally, transport of heat and moisture by the movement of a fluid. In meteorology, the term is used specifically to describe vertical transport of heat and moisture, especially by updrafts and downdrafts in an unstable atmosphere. The terms "convection" and "thunderstorms" often are used interchangeably, although thunderstorms are only one form of convection. Cbs, towering cumulus clouds, and ACCAS clouds all are visible forms of convection. However, convection is not always made visible by clouds. Convection which occurs without cloud formation is called dry convection, while the visible convection processes referred to above are forms of moist convection.

Convergence A contraction of a vector field; the opposite of divergence. Convergence in a horizontal wind field indicates that more air is entering a given area than is leaving at that level. To compensate for the resulting "excess," vertical motion may result: upward forcing if convergence is at low levels, or downward forcing (subsidence) if convergence is at high levels. Upward forcing from low-level convergence increases the potential for thunderstorm development (when other factors, such as instability, are favorable). Compare with confluence.

Cumulus One of the three basic cloud forms (the others are cirrus and stratus). It is also one of the two low cloud types. A cloud that develops in a vertical direction from the base (bottom) up. They have flat bases and dome- or cauliflower-shaped upper surfaces. The base of the cloud is often no more than 3,000 feet above the ground, but the top often varies in height. Small, separate cumulus are associated with fair weather (cumulus humilis). With additional heating from the earth's surface, they can grow vertically throughout the day.

Divergence The expansion or spreading out of a vector field; usually said of horizontal winds. It is the opposite of convergence. Divergence at upper levels of the atmosphere enhances upward motion, and hence the potential for thunderstorm development (if other factors also are favorable).

Dust Bowl The term given to the area of the Great Plains including Texas, Oklahoma, Kansas, Colorado, and New Mexico that was most greatly affected during the Great Drought of the 1930s.

El Niño The cyclical warming of East Pacific Ocean sea water temperatures off the western coast of South America that can result in significant changes in weather patterns in the United States and elsewhere. This occurs when warm equatorial waters move in and displace the colder waters of the Humbolt Current, cutting off the upwelling process.

Flash Flood A flood which is caused by heavy or excessive rainfall in a short period of time, generally less than six hours. Also, at times a dam failure can cause a flash flood, depending on the type of dam and time period during which the break occurs.

Fahrenheit Temperature Scale A temperature scale where water at sea level has a freezing point of +32°F and a boiling point of +212°F. More commonly used in areas that observe the English system of measurement. Created in 1714 by Gabriel Daniel Fahrenheit (1696–1736), a German physicist, who also invented the alcohol and mercury thermometers.

Fog Fog is water droplets suspended in the air at the earth's surface. Fog is often hazardous when the visibility is reduced to ¼ mile or less.

Freeze A freeze is when the surface air temperature is expected to be 32°F or below over a widespread area for a climatologically significant period of time. Use of the term is usually restricted to advective situations or to occasions when wind or other conditions prevent frost. "Killing" may be used during the growing season when the temperature is expected to be low enough for a sufficient duration to kill all but the hardiest herbaceous crops.

Front A boundary or transition zone between two air masses of different density, and thus (usually) of different temperature. A moving front is named according to the advancing air mass, e.g., cold front if colder air is advancing.

Frost Frost describes the formation of thin ice crystals on the ground or other surfaces in the form of scales, needles, feathers, or fans. Frost develops under conditions similar to dew, except the temperatures of the earth's surface and earthbound objects fall below 32°F. As with the term "freeze," this condition is primarily significant during the growing season. If a frost period is sufficiently severe to end the growing season or delay its beginning, it is commonly referred to as a "killing frost." Because frost is primarily an event that occurs as the result of radiational cooling, it frequently occurs with a thermometer level temperature in the mid-30s.

Funnel Cloud A condensation funnel extending from the base of a towering cumulus or Cb, associated with a rotating column of air that is not in contact with the ground (and hence different from a tornado). A condensation funnel is a tornado, not a funnel cloud, if either a) it is in contact with the ground or b) a debris cloud or dust whirl is visible beneath it.

Green Flash A brilliant green coloration of the upper edge of the sun, occasionally seen as the sun's apparent disk is about to set below a clear horizon.

Greenhouse Effect The overall warming of the earth's lower atmosphere primarily due to carbon dioxide and water vapor, which permit the sun's rays to heat the earth, but then restrict some heat-energy from escaping back into space.

Ground Fog Fog created when radiational cooling at the earth's surface lowers the temperature of the air near the ground to or below its initial dew point. Primarily takes place at night or early morning.

Gulf Stream The warm, well-defined, swift, relatively narrow ocean current which exists off the east coast of the United States, beginning near Cape Hatteras. The term also applies to the oceanic system of currents that dominate the western and northern Atlantic Ocean.

High Wind Sustained wind speeds of 40 mph or greater lasting for one hour or longer, or winds of 58 mph or greater for any duration.

Humidity Generally, a measure of the water vapor content of the air. Popularly, it is used synonymously with relative humidity.

Ice Storm An ice storm is used to describe occasions when damaging accumulations of ice are expected during freezing rain situations. Significant accumulations of ice pull down trees and utility lines resulting in loss of power and communication. These accumulations of ice make walking and driving extremely dangerous. Significant ice accumulations are usually accumulations of ¼" or greater.

Inversion Generally, a departure from the usual increase or decrease in an atmospheric property with altitude. Specifically it almost always refers to a temperature inversion, i.e., an increase in temperature with height, or to the layer within which such an increase occurs.

Lake Effect Snow Snow showers that are created when cold dry air passes over a large warmer lake, such as one of the Great Lakes, and picks up moisture and heat.

Landspout [Slang] A tornado that does not arise from organized storm-scale rotation and therefore is not associated with a wall cloud (visually) or a mesocyclone (on radar). Landspouts typically are observed beneath Cbs or towering cumulus clouds (often as no more than a dust whirl), and essentially are the land-based equivalents of waterspouts.

Lenticular Cloud A cloud species which has elements resembling smooth lenses or almonds and is more or less isolated. These clouds are caused by a wave wind pattern created by the mountains. They are also indicative of downstream turbulence on the leeward side of a barrier.

Microburst A convective downdraft with an affected outflow area of less than 2½ miles wide and peak winds lasting less than 5 minutes. Microbursts may induce dangerous horizontal/vertical wind shears, which can adversely affect aircraft performance and cause property damage.

Moisture Convergence A measure of the degree to which moist air is converging into a given area, taking into account the effect of converging winds and moisture advection. Areas of persistent moisture convergence are favored regions for thunderstorm development, if other factors (e.g., instability) are favorable.

National Oceanic and Atmospheric Administration (NOAA) A branch of the U.S. Department of Commerce, it is the parent organization of the National Weather Service. It promotes global environmental stewardship, emphasizing atmospheric and marine resources.

National Weather Service (NWS) A primary branch of the National Oceanic and Atmospheric Administration, it is responsible for all aspects of observing and forecasting atmospheric conditions and their consequences, including severe weather and flood warnings.

Ozone Layer An atmospheric layer that contains a high proportion of oxygen that exists as ozone. It acts as a filtering mechanism against incoming ultraviolet radiation. It is located between the troposphere and the stratosphere, around 9.5 to 12.5 miles (15 to 20 kilometers) above the earth's surface.

Radar Acronym for RAdio Detection And Ranging. An electronic instrument used to detect distant objects and measure their range by how they scatter or reflect radio energy. Precipitation and clouds are detected by measuring the strength of the electromagnetic signal reflected back.

Relative Humidity A dimensionless ratio, expressed in percent, of the amount of atmospheric moisture present relative to the amount that would be present if the air were saturated. Since the latter amount is dependent on temperature, relative humidity is a function of both moisture content and temperature. As such, relative humidity by itself does not directly indicate the actual amount of atmospheric moisture present.

Right Mover A thunderstorm that moves appreciably to the right relative to the main steering winds and to other nearby thunderstorms. Right movers typically are associated with a high potential for severe weather.

Roll Cloud A low, horizontal tube-shaped arcus cloud associated with a thunderstorm gust front (or sometimes with a cold front). Roll clouds are relatively rare; they are completely detached from the thunderstorm base or other cloud features, thus differentiating them from the more familiar shelf clouds. Roll clouds usually appear to be "rolling" about a horizontal axis, but should not be confused with funnel clouds.

Severe Local Storm A convective storm that usually covers a relatively small geographic area, or moves in a narrow path, and is sufficiently intense to threaten life and/or property. Examples include severe thunderstorms with large hail, damaging wind, or tornadoes. Although cloud-to-ground lightning is not a criteria for severe local storms, it is acknowledged to be highly dangerous and a leading cause of deaths, injuries, and damage from thunderstorms. A thunderstorm need not be severe to generate frequent cloud-to-ground lightning.

Severe Thunderstorm A thunderstorm that produces a tornado, winds of at least 58 mph (50 knots), and/or hail at least ¾" in diameter. Structural wind damage may imply the occurrence of a severe thunderstorm. A thunderstorm wind equal to or greater than 40 mph (35 knots) and/or hail of at least ½" is defined as approaching severe.

Shear Variation in wind speed (speed shear) and/or direction (directional shear) over a short distance. Shear usually refers to vertical wind shear, i.e., the change in wind with height, but the term also is used in Doppler radar to describe changes in radial velocity over short horizontal distances.

Shelf Cloud A low, horizontal wedge-shaped arcus cloud, associated with a thunderstorm gust front (or occasionally with a cold front, even in the absence of thunderstorms). Unlike the roll cloud, the shelf cloud is attached to the base of the parent cloud above it (usually a thunderstorm). Rising cloud motion often can be seen in the leading (outer) part of the shelf cloud, while the underside often appears turbulent, boiling, and wind-torn.

Snow Flurries Snow flurries are an intermittent light snowfall of short duration (generally light snow showers) with no measurable accumulation (trace category).

Stratiform Having extensive horizontal development, as opposed to the more vertical development characteristic of convection. Stratiform clouds cover large areas but show relatively little vertical development. Stratiform precipitation, in general, is relatively continuous and uniform in intensity (i.e., steady rain versus rain showers).

Stratocumulus Low-level clouds, existing in a relatively flat layer but having individual elements. Elements often are arranged in rows, bands, or waves. Stratocumulus often reveals the depth of the moist air at low levels, while the speed of the cloud elements can reveal the strength of the low-level jet.

Stratus A low, generally gray cloud layer with a fairly uniform base. Stratus may appear in the form of ragged patches, but otherwise does not exhibit individual cloud elements as do cumulus and stratocumulus clouds. Fog usually is a surface-based form of stratus.

Thermodynamics In general, the relationships between heat and other properties (such as temperature, pressure, density, etc.). In forecast discussions, thermodynamics usually refers to the distribution of temperature and moisture (both vertical and horizontal) as related to the diagnosis of atmospheric instability.

Tornado A violently rotating column of air, usually pendant to a cumulonimbus, with circulation reaching the ground. It nearly always starts as a funnel cloud and may be accompanied by a loud roaring noise. On a local scale, it is the most destructive of all atmospheric phenomena.

Towering Cumulus A large cumulus cloud with great vertical development, usually with a cauliflower-like appearance, but lacking the characteristic anvil of a cumulonimbus.

Troposphere The layer of the atmosphere from the earth's surface up to the tropopause, characterized by decreasing temperature with height (except, perhaps, in thin layers—*see* inversion, cap), vertical wind motion, appreciable water vapor content, and sensible weather (clouds, rain, etc.).

Tsunami An ocean wave with a long period that is formed by an underwater earthquake or landslide, or volcanic eruption. It may travel unnoticed across the ocean for thousands of miles from its point of origin and builds up to great heights over shallower water. Also known as a seismic sea wave and, incorrectly, as a tidal wave.

Updraft A small-scale current of rising air. If the air is sufficiently moist, then the moisture condenses to become a cumulus cloud or an individual tower of a towering cumulus.

Virga Streaks or wisps of precipitation falling from a cloud but evaporating before reaching the ground. In certain cases, shafts of virga may precede a microburst.

Warm Advection Transport of warm air into an area by horizontal winds. Low-level warm advection sometimes is referred to (erroneously) as overrunning. Although the two terms are not properly interchangeable, both imply the presence of lifting in low levels.

Waterspout In general, a tornado occurring over water. Specifically, it normally refers to a small, relatively weak rotating column of air over water beneath a cumulonimbus or towering cumulus cloud. Waterspouts are most common over tropical or subtropical waters.

The exact definition of waterspout is debatable. In most cases the term is reserved for small vortices over water that are not associated with storm-scale rotation (i.e., they are the water-based equivalent of landspouts). But there is sufficient justification for calling virtually any rotating column of air a waterspout if it is in contact with a water surface.

Weather Surveillance Radar (WSR-88D) The newest generation of Doppler radars, the 1988 Doppler weather radar. The radar units, with help from a set of computers, show very detailed images of precipitation and other phenomena, including air motions within a storm.

Wind Chill Increased wind speeds accelerate heat loss from exposed skin. No specific rules exist for determining when wind chill becomes dangerous. As a general rule, the threshold for potentially dangerous wind chill conditions is about –20°F.

EXPLANATION OF CLIMATIC DATA CHARTS

The following tables for each month contain comparative information for cities around the world. U.S. climatological information contained within has been furnished by the National Oceanic and Atmospheric Administration. Records of foreign data come from various sources with various lengths of record. The number of years used to determine the values for each category in the tables varies, depending in part on how long the various information has been recorded.

RECORDS *(Temperature & Precipitation)*

Records for individual cities cover extremes observed during the years observations have been taken. For cities such as New York, where several observation sites are located, the record given is the extreme highest or lowest reading obtained from all locations.

AVERAGES *(Temperature & Precipitation)*

Monthly normals are generally 30-year average values. Mathematical adjustments have been made to render the data representative of the current location. The values have been statistically determined and cannot be recreated solely from the original record.

Average Max: The average daily maximum temperature, °F.

Average Min: The average daily minimum temperature, °F.

Record Max: Highest ever recorded for month.

Record Min: Lowest ever recorded for month.

Mean Number of Days of Precipitation: Mean number of days for month with .01" or more precipitation.

Amount of Precipitation: *(Snowfall & Rainfall)* Average monthly and extreme monthly rainfall and snowfall totals.

Percentage of Sunshine: Average percentage of possible sunshine for month.

Mean Number of Clear Days: Number of days with average cloud cover of less than 3/10.

Mean Number of Cloudy Days: Number of days with 8/10 or more average cloud cover.

Relative Humidity: Average for warmest time of day.

Average Wind Speed: Average speed of the wind regardless of direction.

Degree Days: Degree day data are used to estimate amounts of energy required to maintain comfortable indoor temperature levels. Daily values are computed from each day's mean temperature ([max plus min]/2). Each degree that a day's mean temperature is below or above 65°F is counted as one heating or cooling degree day.

"T" stands for trace. "NA" stands for data not available.

Climatic data is current as of 2007.

U.S. & INTERNATIONAL CLIMATE DATA

JANUARY—U.S. CLIMATIC DATA

	TEMPERATURE (°F) AVERAGE MIN	TEMPERATURE (°F) AVERAGE MAX	TEMPERATURE (°F) RECORD MIN	TEMPERATURE (°F) RECORD MAX	MEAN DAYS OF PRECIP	AMOUNT OF PRECIPITATION TOTAL AVG	AMOUNT OF PRECIPITATION TOTAL MAX	AMOUNT OF PRECIPITATION SNOWFALL AVG	AMOUNT OF PRECIPITATION SNOWFALL MAX	% OF POSSIBLE SUN	MEAN CLEAR DAYS	MEAN CLOUDY DAYS	REL HUMIDITY	AVG WIND SPEED (MPH)	DEGREE DAYS (BASE 65°F) HEATING	DEGREE DAYS (BASE 65°F) COOLING
Albany, NY	11.0	30.2	-28	71	12.4	2.36	7.30	16.0	47.8	45	5.5	17.4	71	9.8	1376	0
Albuquerque, NM	21.7	46.8	-17	69	3.8	0.44	2.52	2.5	9.5	72	13.0	10.2	40	8.1	955	0
Amarillo, TX	21.2	49.0	-11	83	4.0	0.50	2.67	0.0	15.9	69	12.4	11.2	50	13.0	927	0
Anchorage, AK	8.4	21.4	-35	56	7.0	0.79	2.71	10.1	36.1	40	8.3	18.2	70	6.1	1553	0
Atlanta, GA	31.5	50.4	-8	79	11.6	4.75	15.82	0.8	10.3	49	8.3	16.4	60	10.6	744	0
Atlantic City, NJ	21.4	40.4	-10	78	10.8	3.46	8.40	5.0	22.3	49	8.4	14.7	70	11.3	1057	0
Bakersfield, CA	38.6	56.9	14	82	5.9	0.86	3.90	T	3.0	NA	7.0	16.0	63	5.2	533	0
Bismarck, ND	-1.7	20.2	-45	63	7.9	0.45	1.64	7.2	25.0	54	6.8	16.5	67	10.0	1730	0
Boise, ID	21.6	36.4	-28	63	12.4	1.45	5.29	7.2	27.0	39	4.4	21.8	70	8.2	1116	0
Boston, MA	21.6	35.7	-13	72	11.7	3.59	10.55	12.6	43.3	53	9.3	15.0	57	14.0	1128	0
Buffalo, NY	17.0	30.2	-16	72	20.2	2.70	6.88	23.9	68.3	32	1.4	23.4	73	14.3	1283	0
Burlington, VT	7.5	25.1	-30	66	14.3	1.82	5.15	18.5	42.4	41	4.3	20.1	66	9.5	1510	0
Caribou, ME	-1.6	19.4	-33	53	14.4	2.42	5.68	23.8	44.5	NA	6.8	17.3	71	12.4	1739	0
Casper, WY	12.0	32.8	-40	60	7.4	0.55	1.42	10.1	39.3	NA	6.5	16.8	60	16.5	1321	0
Charleston, SC	37.7	57.8	6	83	9.8	3.45	8.92	0.1	1.9	58	8.9	15.6	55	9.2	548	15
Charleston, WV	23.0	41.2	-16	81	15.6	2.91	9.11	10.4	39.5	NA	3.5	20.7	65	7.6	1020	0
Charlotte, NC	29.6	49.0	-5	79	10.2	3.71	10.39	2.0	13.5	55	9.1	15.5	56	7.9	797	0
Chicago, IL	12.9	29.0	-27	67	11.3	1.53	4.84	11.1	42.5	42	6.9	18.2	67	11.6	1364	0
Cincinnati, OH	19.5	36.6	-25	77	12.1	2.59	13.68	7.5	31.5	41	5.1	19.7	68	10.7	1144	0
Cleveland, OH	17.6	31.9	-20	73	16.5	2.04	7.01	16.6	42.8	31	27.0	23.5	70	12.4	1246	0
Dallas, TX	32.7	54.1	-3	93	7.0	1.83	9.07	1.4	12.1	53	9.8	15.5	59	11.1	670	0
Denver, CO	16.1	43.2	-29	76	5.8	0.50	2.35	7.7	23.7	71	10.0	11.8	46	8.8	1094	0
Des Moines, IA	10.7	28.1	-30	67	7.6	0.96	4.38	8.6	37.0	51	7.8	16.2	68	11.7	1414	0
Detroit, MI	15.6	30.3	-21	67	13.0	1.76	5.02	9.6	29.6	40	4.3	20.0	70	11.7	1305	0
El Paso, TX	29.4	56.1	-8	80	4.1	0.40	1.84	1.3	8.3	77	13.9	9.8	36	8.6	688	0
Ely, NV	9.4	39.7	-27	68	6.8	0.70	2.08	24.8	—	67	8.6	14.9	55	10.2	1252	0
Grand Rapids, MI	14.7	29.0	-22	66	16.3	1.83	6.00	24.1	46.8	30	2.5	23.6	72	11.4	1339	0
Hartford, CT	15.8	33.2	-26	70	10.7	3.41	9.61	12.1	43.1	57	7.7	15.0	61	9.0	1252	0
Helena, MT	9.6	29.6	-42	63	8.2	0.63	3.75	9.1	35.6	46	4.6	20.5	61	6.9	1407	0
Honolulu, HI	65.6	80.1	53	88	10.1	3.55	18.36	0.0	—	62	9.1	8.9	63	9.8	0	245
Houston, TX	39.7	61.0	5	84	10.3	3.29	13.11	0.2	3.0	43	7.3	18.3	65	8.2	468	16
Indianapolis, IN	17.2	33.7	-27	71	11.9	2.32	12.69	6.2	30.6	41	5.9	19.1	72	11.0	1225	0
Jackson, MS	32.7	55.6	-5	85	11.0	5.24	14.10	.07	11.7	47	7.8	17.2	71	8.7	656	8
Jacksonville, FL	40.5	64.2	7	85	7.8	3.31	10.21	T	T	58	9.2	13.5	74	8.3	421	31
Kansas City, MO	16.7	34.7	-20	75	8.2	1.09	5.52	5.8	30.5	59	9.6	15.4	64	11.0	1218	0
Las Vegas, NV	33.6	57.3	8	77	3.0	0.48	3.00	1.2	16.7	77	13.9	10.9	30	7.4	605	0
Lexington, KY	22.4	39.1	-21	80	12.6	2.86	16.65	6.2	21.9	NA	5.8	19.3	69	11.2	1060	0
Lincoln, NE	10.1	32.4	-33	73	6.5	0.54	3.70	6.5	23.0	57	9.2	15.6	70	10.1	1355	0
Little Rock, AR	29.4	47.6	-8	83	9.7	3.21	18.04	2.3	19.4	46	8.6	16.3	64	8.7	822	0
Los Angeles, CA	48.9	67.7	23	95	6.1	2.92	14.94	T	2.0	69	12.0	10.8	56	6.7	222	14
Lubbock, TX	24.6	52.9	-16	87	3.7	0.39	4.05	2.5	25.3	66	12.6	11.9	47	12.1	812	0
Marquette, MI	3.3	20.5	-33	57	17.5	2.17	6.61	26.7	91.7	32	NA	NA	NA	NA	1646	0
Medford, OR	30.4	45.7	-3	71	13.5	2.69	6.67	3.6	22.6	NA	2.6	23.5	72	4.1	834	0
Memphis, TN	30.9	48.5	-8	79	10.1	3.73	17.56	2.3	15.1	50	7.9	17.2	63	10.3	784	0
Miami, FL	59.2	75.2	30	88	6.5	2.01	7.93	0.0	—	67	9.6	8.5	60	9.4	88	156
Milwaukee, WI	11.6	26.1	-26	62	11.1	1.60	5.38	13.0	52.6	44	7.2	17.7	68	12.8	1429	0
Minneapolis, MN	2.8	20.7	-34	58	8.8	0.95	3.63	9.9	46.4	52	8.4	15.3	66	10.4	1649	0
Montgomery, AL	35.8	56.3	0	83	10.7	4.68	17.78	0.2	6.0	48	7.4	17.1	64	7.8	594	8
Nashville, TN	26.5	45.9	-17	78	11.1	3.58	14.75	4.2	18.8	41	6.3	18.4	63	9.2	893	0
New Orleans, LA	41.8	60.8	14	83	10.1	5.05	19.28	0.0	5.0	48	6.7	16.9	66	9.4	450	25
New York, NY	25.3	37.6	-6	72	11.2	3.42	10.52	7.6	27.4	50	8.1	13.7	60	10.7	1039	0
Norfolk, VA	30.9	47.3	-7	80	10.4	3.78	9.93	2.9	22.7	56	9.2	15.4	67	11.5	803	0
Oklahoma City, OK	25.2	46.7	-11	83	5.5	1.13	5.68	2.9	17.3	59	10.3	14.6	60	12.9	902	0
Orlando, FL	48.6	70.8	19	87	6.3	2.30	7.23	T	T	NA	9.4	11.1	56	9.0	234	70
Philadelphia, PA	22.8	37.9	-7	74	11.0	3.21	8.86	6.4	33.7	50	7.4	16.1	59	10.3	1073	0
Phoenix, AZ	39.4	65.2	16	88	3.8	.73	5.25	T	1.0	78	13.8	10.3	32	5.3	394	0
Pittsburgh, PA	18.5	33.7	-22	75	16.4	2.54	7.15	12.1	40.2	33	3.0	21.9	65	10.7	1206	0
Portland, OR	33.7	45.4	-2	66	18.6	5.35	13.71	3.9	41.4	27	2.8	25.0	76	10.0	787	0
Providence, RI	19.1	36.6	-13	69	11.1	3.88	11.66	9.9	37.4	57	9.9	14.4	63	11.3	1150	0
Raleigh, NC	28.8	48.9	-9	80	9.9	3.48	7.52	0.0	25.8	54	9.3	14.7	55	8.6	809	0
Reno, NV	20.7	45.1	-19	71	6.1	1.07	6.76	5.9	65.7	65	8.5	15.4	51	5.6	995	0
Sacramento, CA	37.7	52.7	19	72	10.2	3.73	15.04	T	4.0	45	6.3	18.7	71	7.6	614	0
Salt Lake City, UT	19.3	36.4	-22	63	9.9	1.11	3.90	13.2	33.6	46	5.5	18.9	68	7.7	1150	0
San Antonio, TX	37.9	60.8	0	89	8.1	1.71	8.52	0.2	15.9	48	8.9	15.9	57	9.2	494	8
San Diego, CA	48.9	65.9	25	88	6.9	1.80	9.09	T	T	72	12.4	11.2	56	5.8	245	9
San Francisco, CA	45.8	56.3	24	79	10.8	4.06	24.36	0.0	1.5	56	8.6	14.7	66	7.1	431	0
Seattle, WA	35.2	45.0	0	67	19.0	5.38	12.92	6.0	57.2	24	2.6	24.5	74	9.9	772	0
Sioux Falls, SD	3.3	24.3	-38	66	6.2	0.51	2.23	6.5	22.2	NA	8.1	15.2	71	11.0	1587	0
Spokane, WA	20.8	33.2	-30	62	14.2	1.98	4.96	17.3	56.9	26	3.2	23.5	78	8.6	1175	0
St. Louis, MO	20.8	37.7	-22	77	8.6	1.81	9.00	5.4	23.9	52	7.5	17.1	71	10.6	1107	0
Tampa, FL	50.0	69.8	21	86	6.3	1.99	8.02	0.0	0.2	64	9.5	11.6	59	8.8	234	76
Washington, DC	26.8	42.3	-18	79	10.3	2.72	7.83	5.1	31.5	48	7.6	16.2	54	10.0	942	0
Wichita, KS	19.2	39.8	-15	75	5.6	0.79	6.29	4.5	19.7	59	10.1	14.8	65	12.2	1101	0

INTERNATIONAL DATA

	AVG HIGH TEMP	AVG LOW TEMP	AVG PRECIP	DAYS WITH MEASURABLE PRECIP
Cape Town, South Africa	78	60	0.7	NA
Delhi, India	70	44	0.9	NA
Hong Kong, China	64	56	1.3	NA
Jerusalem, Israel	55	41	5.2	NA
London, United Kingdom	43	36	2.1	15
Manila, Philippines	87	68	0.6	4
Melbourne, Australia	78	57	1.9	8
Montreal, Canada	22	6	2.8	17
Moscow, Russia	21	12	1.2	16
Nairobi, Kenya	80	55	1.9	5
Paris, France	43	34	2.2	17
Rio de Janeiro, Brazil	84	73	4.9	13
Rome, Italy	52	40	2.8	8
Sydney, Australia	78	65	3.9	13
Tahiti, Society Islands	89	72	9.9	NA
Tokyo, Japan	47	29	1.9	6
Toronto, Canada	28	12	2.0	14

FEBRUARY—U.S. CLIMATIC DATA

	TEMPERATURE (°F) AVERAGE MIN	AVERAGE MAX	RECORD MIN	RECORD MAX	MEAN DAYS OF PRECIP	AMOUNT OF PRECIPITATION TOTAL AVG	TOTAL MAX	SNOWFALL AVG	SNOWFALL MAX	% OF POSSIBLE SUN	MEAN CLEAR DAYS	MEAN CLOUDY DAYS	REL HUMID-ITY	AVG WIND SPEED (MPH)	DEGREE DAYS (BASE 65°F) HEATING	COOLING
Albany, NY	13.8	33.2	-22	68	10.6	2.27	5.19	14.3	40.7	51	5.4	15.4	66	10.3	1162	0
Albuquerque, NM	26.4	53.5	-6	75	4.0	0.46	2.60	2.0	10.3	73	11.3	9.2	32	8.9	700	0
Amarillo, TX	25.5	52.8	-16	88	4.4	0.61	2.93	0.0	28.7	69	10.4	10.0	49	14.1	722	0
Anchorage, AK	11.5	25.8	-38	57	8.3	0.78	3.07	11.7	52.1	44	6.6	18.0	66	6.7	1296	0
Atlanta, GA	34.5	55.0	-9	80	10.1	4.81	12.77	0.5	11.6	54	8.0	13.8	54	10.9	566	0
Atlantic City, NJ	23.5	42.5	-11	77	9.7	3.06	7.44	5.5	35.2	52	7.7	13.6	68	11.6	896	0
Bakersfield, CA	42.6	63.9	20	85	6.2	1.06	5.36	T	T	NA	7.6	12.1	51	5.8	331	0
Bismarck, ND	5.1	26.4	-45	69	6.8	0.43	1.74	6.4	25.6	54	5.2	15.3	67	9.9	1380	0
Boise, ID	27.5	44.2	-15	71	10.5	1.07	6.49	3.7	25.2	50	4.3	17.6	60	9.2	815	0
Boston, MA	23.0	37.5	-18	70	10.6	3.62	7.81	11.6	41.6	56	8.4	13.2	56	13.9	972	0
Buffalo, NY	17.4	31.6	-20	71	16.9	2.31	5.90	17.9	54.2	38	2.0	20.7	70	13.7	1134	0
Burlington, VT	8.9	27.5	-30	62	11.6	1.63	5.38	16.7	34.3	47	4.4	17.5	65	9.2	1310	0
Caribou, ME	0.7	23.0	-41	59	12.5	1.92	4.13	21.9	41.0	NA	5.9	16.1	69	12.0	1487	0
Casper, WY	16.0	37.0	-32	68	7.9	0.60	1.42	9.7	23.8	NA	6.1	13.9	56	15.2	1078	0
Charleston, SC	40.0	61.0	7	87	8.9	3.30	10.45	0.3	7.1	62	9.1	12.6	51	10.0	414	8
Charleston, WV	25.7	45.3	-12	80	13.6	3.04	8.10	8.5	21.8	NA	4.3	17.8	61	7.6	826	0
Charlotte, NC	31.9	53.0	-5	82	9.6	3.84	8.58	1.9	17.4	60	8.9	13.0	52	8.4	630	0
Chicago, IL	17.2	33.5	-21	71	9.6	1.36	5.98	7.8	27.8	44	6.1	15.9	65	11.4	1109	0
Cincinnati, OH	22.7	40.8	-17	76	11.1	2.68	8.87	5.3	21.4	51	5.2	17.4	64	10.4	930	0
Cleveland, OH	19.3	35.0	-16	74	14.4	2.19	7.73	13.8	30.5	37	3.0	19.8	68	12.1	1058	0
Dallas, TX	36.9	58.9	2	95	6.4	2.18	7.68	1.1	13.5	58	9.9	12.5	53	11.9	484	5
Denver, CO	20.2	46.6	-30	77	5.7	0.57	2.01	7.4	22.1	71	8.2	11.4	42	9.1	885	0
Des Moines, IA	15.6	33.7	-26	78	7.2	1.11	3.20	7.2	21.3	54	7.5	14.9	66	11.6	1128	0
Detroit, MI	17.6	33.3	-20	70	10.9	1.74	6.41	8.1	38.4	47	4.9	16.5	65	11.4	1109	0
El Paso, TX	33.9	62.2	5	86	2.8	0.41	1.92	0.7	8.9	82	14.0	6.8	26	9.4	473	0
Ely, NV	15.4	43.6	-30	67	7.1	0.65	2.19	0.0	20.0	67	6.9	14.4	50	10.4	994	0
Grand Rapids, MI	15.8	31.6	-24	69	11.9	1.42	7.87	11.2	35.5	39	3.5	18.9	68	10.5	1156	0
Hartford, CT	18.6	36.4	-24	73	10.4	3.23	7.27	12.1	32.7	58	6.5	13.8	60	9.4	1050	0
Helena, MT	15.9	36.9	-42	69	6.4	0.41	1.69	6.0	31.1	54	4.0	17.7	54	7.5	1081	0
Honolulu, HI	65.4	80.5	53	88	9.4	2.21	13.68	0.0	—	64	7.5	8.1	59	10.5	0	224
Houston, TX	42.6	65.3	6	91	7.3	2.96	9.01	0.2	20.0	51	7.7	14.3	59	8.8	322	11
Indianapolis, IN	20.9	38.3	-21	76	10.1	2.46	7.28	5.8	21.7	50	5.7	16.1	70	10.8	991	0
Jackson, MS	35.7	60.1	10	89	8.9	4.70	12.94	0.2	9.1	55	8.5	13.5	63	8.7	485	6
Jacksonville, FL	43.3	67.0	10	88	7.8	3.93	11.12	T	1.9	62	8.9	11.9	68	9.2	296	22
Kansas City, MO	21.8	40.6	-22	81	6.6	1.10	6.76	4.3	20.7	56	7.8	14.1	62	11.6	946	0
Las Vegas, NV	38.8	63.3	-17	87	2.6	0.48	2.89	0.0	4.1	80	12.5	8.7	26	8.5	389	0
Lexington, KY	25.3	43.6	-20	80	11.2	3.21	11.06	4.8	17.4	NA	6.1	16.5	64	11.1	854	0
Lincoln, NE	15.1	37.9	-26	84	5.2	0.72	3.06	6.2	26.1	57	8.0	13.9	67	10.5	1075	0
Little Rock, AR	33.6	52.7	-12	87	8.9	3.53	12.74	1.5	15.6	54	9.1	13.5	59	9.1	610	0
Los Angeles, CA	50.6	69.4	28	95	6.1	3.07	13.68	T	T	72	11.3	10.7	59	7.4	170	32
Lubbock, TX	28.6	57.6	-17	89	4.0	0.68	5.83	3.2	16.8	67	11.0	9.9	41	13.4	613	0
Marquette, MI	4.0	23.9	-34	69	13.2	1.73	5.35	22.3	54.3	37	NA	NA	NA	NA	1428	0
Medford, OR	32.2	53.3	6	79	11.4	1.93	5.67	1.3	17.5	NA	3.4	19.1	59	4.6	622	0
Memphis, TN	34.8	53.5	-11	81	9.5	4.35	10.50	1.3	10.3	54	7.9	14.7	59	10.3	582	0
Miami, FL	60.4	76.5	27	89	6.1	2.08	8.07	0.0	—	65	8.5	7.9	57	10.1	51	149
Milwaukee, WI	15.9	30.1	-26	68	9.5	1.45	5.39	9.3	42.0	47	6.6	15.7	67	12.6	1176	0
Minneapolis, MN	9.2	26.6	-33	64	7.5	0.88	3.25	8.4	26.5	58	7.8	13.6	65	10.5	1319	0
Montgomery, AL	38.8	60.8	-5	85	9.3	5.48	13.38	0.1	4.1	54	8.0	13.8	57	8.3	426	0
Nashville, TN	29.9	50.8	-13	84	10.7	3.81	12.37	3.2	18.9	47	6.9	15.4	59	9.4	689	0
New Orleans, LA	44.4	64.1	7	85	9.1	6.01	13.85	0.1	8.2	53	7.9	13.7	63	9.8	316	17
New York, NY	26.9	40.3	-15	75	9.8	3.27	6.87	8.6	27.9	55	8.3	11.2	58	10.8	879	0
Norfolk, VA	32.3	49.7	-5	82	10.0	3.47	8.21	2.8	18.9	59	8.6	13.7	65	11.9	672	0
Oklahoma City, OK	29.6	52.1	-17	92	6.3	1.56	4.63	2.5	12.9	60	9.1	12.2	54	13.3	675	0
Orlando, FL	49.7	72.7	28	90	7.0	3.02	8.32	0.0	—	NA	8.8	10.7	53	9.6	164	58
Philadelphia, PA	24.8	41.0	-11	79	9.3	2.79	6.87	6.7	31.5	53	7.2	13.7	56	11.0	896	0
Phoenix, AZ	42.5	69.7	22	92	3.9	0.59	4.64	0.0	0.6	80	12.6	89.0	27	5.9	269	20
Pittsburgh, PA	20.3	36.9	-20	77	14.1	2.39	6.52	10.0	25.3	38	3.3	19.0	62	10.6	1016	0
Portland, OR	36.1	51.0	-3	71	16.5	3.85	13.36	0.6	20.0	37	2.6	22.2	68	9.2	599	0
Providence, RI	20.9	38.3	-17	72	10.1	3.61	7.20	10.0	30.9	57	7.9	12.8	61	11.6	988	0
Raleigh, NC	31.3	52.6	-2	84	9.7	3.69	9.73	0.0	17.9	58	8.8	13.2	51	8.9	644	0
Reno, NV	24.2	51.7	-12	76	5.9	0.99	4.99	4.7	32.5	68	6.8	14.0	40	6.2	756	0
Sacramento, CA	41.4	60.0	21	80	8.8	2.87	9.25	0.1	2.0	61	7.5	13.8	61	7.7	400	0
Salt Lake City, UT	24.6	43.6	-30	69	8.8	1.23	4.89	9.5	30.0	55	5.2	16.0	59	8.2	865	0
San Antonio, TX	41.3	65.7	4	100	7.7	1.81	7.88	0.2	4.2	53	8.4	13.8	52	9.8	332	10
San Diego, CA	50.7	66.5	34	90	6.0	1.53	9.05	0.0	—	72	10.6	10.2	58	6.4	189	10
San Francisco, CA	48.7	60.0	25	81	9.8	2.95	14.89	T	3.7	62	7.8	13.1	65	8.5	297	0
Seattle, WA	37.4	49.5	1	74	16.2	3.99	9.11	1.4	35.4	37	2.5	21.7	67	9.7	602	0
Sioux Falls, SD	9.7	29.6	-42	70	6.5	0.64	4.05	8.5	48.4	NA	6.9	14.6	70	11.1	1268	0
Spokane, WA	25.9	40.6	-24	63	11.7	1.49	5.62	7.3	37.8	38	3.2	20.1	69	9.1	888	0
St. Louis, MO	25.1	42.6	-18	85	8.2	2.12	8.94	4.5	23.5	53	7.0	14.8	66	10.9	871	0
Tampa, FL	51.6	71.4	22	88	6.9	3.08	10.85	T	0.1	66	9.1	10.2	56	9.4	160	62
Washington, DC	29.1	45.9	-15	84	8.7	2.71	6.84	5.6	35.2	52	7.6	14.1	52	10.3	770	0
Wichita, KS	23.7	45.9	-22	87	5.2	0.96	4.61	4.4	20.5	60	8.3	12.8	60	12.8	846	0

INTERNATIONAL DATA

	AVG HIGH TEMP	AVG LOW TEMP	AVG PRECIP	DAYS WITH MEASURABLE PRECIP
Cape Town, South Africa	79	60	0.6	NA
Delhi, India	75	49	0.7	NA
Hong Kong, China	63	55	1.8	NA
Jerusalem, Israel	56	42	5.2	NA
London, United Kingdom	44	36	1.6	13
Manila, Philippines	89	69	0.3	3
Melbourne, Australia	78	58	2.0	7
Montreal, Canada	24	8	2.6	14
Moscow, Russia	22	12	1.1	12
Nairobi, Kenya	82	56	1.4	4
Paris, France	45	34	1.8	14
Rio de Janeiro, Brazil	85	73	4.8	11
Rome, Italy	55	42	2.4	9
Sydney, Australia	78	65	4.5	13
Tahiti, Society Islands	89	72	9.6	NA
Tokyo, Japan	48	31	2.9	7
Toronto, Canada	29	13	1.8	12

MARCH—U.S. CLIMATIC DATA

	TEMPERATURE (°F)				MEAN DAYS OF PRECIP	AMOUNT OF PRECIPITATION				% OF POSSIBLE SUN	MEAN		REL HUMID-ITY	AVG WIND SPEED (MPH)	DEGREE DAYS (BASE 65°F)	
	AVERAGE		RECORD			TOTAL		SNOWFALL								
	MIN	MAX	MIN	MAX		AVG	MAX	AVG	MAX		CLEAR DAYS	CLOUDY DAYS			HEATING	COOLING
Albany, NY	24.5	44.0	-21	89	12.1	2.93	7.37	11.9	50.9	53	6.0	17.1	61	10.7	952	0
Albuquerque, NM	32.2	61.4	8	85	4.5	0.54	2.18	1.9	13.9	73	11.3	9.6	24	10.2	561	0
Amarillo, TX	32.7	61.6	-3	96	4.6	0.96	4.14	2.4	21.5	71	11.4	11.0	42	15.5	555	0
Anchorage, AK	18.1	33.1	-24	56	7.7	0.69	2.76	9.1	31.0	53	7.4	17.9	56	6.7	1218	0
Atlanta, GA	42.5	64.3	8	89	11.7	5.77	13.28	0.4	7.9	58	8.9	14.9	51	10.9	365	8
Atlantic City, NJ	31.3	51.6	5	87	10.8	3.62	8.82	2.9	23.6	53	7.7	15.2	65	12.1	729	0
Bakersfield, CA	45.8	68.9	21	94	6.6	1.04	4.61	0.0	1.5	NA	10.1	11.4	43	6.5	246	11
Bismarck, ND	17.8	38.5	-36	81	8.2	0.77	3.27	8.4	31.1	59	5.4	17.2	63	11.0	1141	0
Boise, ID	31.9	52.9	5	81	9.7	1.29	7.66	1.9	18.9	62	6.0	17.8	45	10.1	701	0
Boston, MA	31.3	45.8	-8	89	11.9	3.69	11.00	7.8	38.9	57	7.8	15.3	57	13.8	818	0
Buffalo, NY	25.9	41.7	-7	81	16.2	2.68	7.03	11.8	38.5	45	3.6	19.9	67	13.4	967	0
Burlington, VT	22.0	39.3	-24	84	13.1	2.23	4.53	12.4	47.6	50	5.6	18.6	63	9.4	1063	0
Caribou, ME	14.9	34.3	-28	73	12.9	2.43	5.13	20.4	47.1	NA	6.5	17.0	68	12.9	1252	0
Casper, WY	21.8	45.2	-21	74	9.6	0.95	2.43	15.2	36.2	NA	5.8	16.5	48	14.0	977	0
Charleston, SC	47.5	68.6	15	94	10.4	4.34	11.11	0.1	2.0	67	9.1	13.7	51	10.1	239	25
Charleston, WV	35.0	56.7	0	92	15.2	3.63	8.94	4.9	21.4	NA	4.4	18.8	54	8.4	592	0
Charlotte, NC	39.4	62.3	4	91	11.4	4.43	11.13	1.5	19.3	63	9.1	13.8	51	8.9	437	0
Chicago, IL	28.5	45.8	-12	88	12.7	2.69	5.91	7.6	24.7	50	4.7	17.6	61	11.8	862	0
Cincinnati, OH	33.1	53.0	-11	88	13.2	4.24	12.8	4.5	13.9	28	5.0	19.3	61	11.1	682	0
Cleveland, OH	28.2	46.3	-5	83	15.6	2.91	8.31	9.9	26.3	44	4.4	20.2	63	12.4	859	0
Dallas, TX	45.6	67.8	11	96	7.3	2.77	9.53	0.2	4.3	59	9.6	13.7	51	13.0	286	29
Denver, CO	25.8	52.2	-11	84	8.7	1.28	4.56	13.1	35.2	70	7.8	12.9	41	9.8	806	0
Des Moines, IA	27.6	46.9	-22	91	10.2	2.33	5.82	7.0	28.0	54	6.6	17.1	60	12.9	859	0
Detroit, MI	27.0	44.4	-4	82	13.2	2.55	5.63	6.8	30.2	51	5.3	18.3	62	11.7	908	0
El Paso, TX	40.2	69.9	14	93	2.4	0.29	2.26	0.5	7.3	85	14.8	8.1	21	11.4	316	10
Ely, NV	20.7	48.4	-13	76	8.5	0.96	2.40	9.9	24.8	70	7.6	15.2	43	10.7	942	0
Grand Rapids, MI	25.4	42.8	-13	82	12.8	2.63	6.88	11.0	36.0	44	4.5	19.6	64	11.1	958	0
Hartford, CT	28.1	46.8	-6	89	11.5	3.63	9.21	10.9	43.3	56	6.4	16.2	57	10.0	853	0
Helena, MT	22.3	44.8	-30	77	8.6	0.73	2.39	7.4	26.6	50	3.6	19.1	46	8.5	973	0
Honolulu, HI	67.2	81.6	55	88	9.0	2.20	20.79	0.0	—	68	7.3	9.6	58	11.5	0	291
Houston, TX	50.0	71.1	21	96	9.8	2.92	10.66	0.0	0.8	46	6.5	17.9	50	9.5	187	50
Indianapolis, IN	31.9	50.9	-7	85	13.1	3.79	10.95	3.7	30.4	50	5.6	18.3	65	11.7	732	0
Jackson, MS	44.1	69.3	15	95	10.5	5.82	15.13	0.3	5.3	59	8.5	15.1	59	9.3	285	28
Jacksonville, FL	49.2	73.0	23	91	8.0	3.68	12.52	T	T	67	9.2	12.5	65	9.2	169	48
Kansas City, MO	32.6	52.8	-10	91	10.7	2.51	9.08	3.9	40.2	58	6.3	16.9	59	12.5	691	0
Las Vegas, NV	43.8	68.8	19	92	2.9	0.42	4.80	0.0	0.1	83	13.8	8.2	22	10.1	292	22
Lexington, KY	35.3	55.3	-2	86	13.1	4.40	13.82	2.8	17.7	NA	5.5	18.0	58	11.5	611	0
Lincoln, NE	26.8	50.3	-19	91	8.8	2.09	6.65	6.5	21.3	56	7.8	15.7	59	11.9	818	0
Little Rock, AR	42.6	62.7	11	91	10.4	4.97	10.43	0.6	8.0	57	8.6	15.4	55	9.8	395	13
Los Angeles, CA	51.8	69.5	31	99	5.8	2.61	12.36	0.0	—	73	11.6	10.7	61	8.1	169	36
Lubbock, TX	36.4	66.0	-2	95	3.7	0.89	5.94	1.6	16.5	72	11.6	0.6	33	13.4	437	10
Marquette, MI	13.6	33.5	-30	71	14.8	2.77	6.08	21.2	60.6	44	NA	NA	NA	NA	1283	0
Medford, OR	35.4	58.5	14	86	11.7	1.82	5.54	0.8	8.1	NA	5.2	19.3	50	5.3	558	0
Memphis, TN	43.0	63.2	12	87	11.0	5.41	13.04	1.0	18.5	56	8.1	16.3	56	11.1	383	14
Miami, FL	64.2	79.1	32	93	5.7	2.39	10.57	0.0	—	77	8.7	8.1	56	10.4	14	221
Milwaukee, WI	26.2	40.4	-10	81	11.9	2.67	6.93	9.2	36.3	50	6.0	17.2	65	13.0	983	0
Minneapolis, MN	22.7	39.2	-32	83	10.3	1.94	4.75	10.6	40.0	55	6.9	16.4	62	11.3	1054	0
Montgomery, AL	45.7	68.6	17	90	10.3	6.26	16.51	T	1.4	59	8.0	15.1	55	8.4	263	21
Nashville, TN	39.1	61.2	2	89	12.1	4.85	12.35	1.5	21.5	52	7.6	16.6	53	10.0	469	10
New Orleans, LA	51.6	71.6	25	90	9.0	4.90	21.09	T	T	57	7.7	15.1	60	9.9	162	56
New York, NY	34.8	50.0	3	86	11.5	4.08	10.41	5.1	30.5	56	8.8	12.1	55	11.0	701	0
Norfolk, VA	39.3	57.9	8	92	11.0	3.70	9.01	1.2	19.1	63	9.0	14.6	62	12.4	508	0
Oklahoma City, OK	38.5	62.0	1	97	7.2	2.71	8.02	1.5	20.7	63	9.6	13.4	49	14.6	464	9
Orlando, FL	55.2	78.0	25	92	7.5	3.21	11.38	0.0	—	NA	9.3	11.2	50	9.9	65	117
Philadelphia, PA	33.2	51.6	5	87	11.1	3.46	9.10	4.0	15.2	55	7.5	15.3	53	11.4	701	0
Phoenix, AZ	46.7	74.5	25	100	3.5	0.81	4.82	T	0.2	83	14.6	8.5	24	6.7	187	51
Pittsburgh, PA	29.8	49.0	-1	84	16.1	3.41	6.10	8.7	34.1	44	4.0	20.1	58	10.9	794	0
Portland, OR	38.6	56.0	19	83	17.2	3.56	12.76	0.5	15.2	46	3.1	23.6	60	8.3	549	0
Providence, RI	28.8	46.1	1	90	11.8	4.05	8.84	8.3	31.6	57	8.6	14.9	60	12.2	856	0
Raleigh, NC	38.7	62.1	11	94	10.2	3.77	7.78	1.5	17.8	62	9.6	13.8	49	9.4	458	6
Reno, NV	29.2	56.3	-3	83	6.4	0.71	4.15	4.9	29.0	75	8.2	13.6	34	7.7	688	0
Sacramento, CA	43.2	64.0	26	88	8.4	2.57	10.00	T	2.0	72	10.4	12.5	53	8.8	357	0
Salt Lake City, UT	31.4	52.2	0	78	9.9	1.91	4.66	10.4	41.9	63	7.0	15.7	47	9.3	719	0
San Antonio, TX	49.7	73.5	19	100	7.0	1.52	7.24	T	1.5	57	8.5	15.2	47	10.6	167	64
San Diego, CA	52.8	66.3	36	99	7.2	1.77	7.20	0.0	—	71	11.2	10.5	59	7.3	177	9
San Francisco, CA	49.0	60.8	30	87	9.7	3.07	9.04	T	1.0	69	9.7	12.6	63	10.4	313	0
Seattle, WA	38.5	52.7	11	81	17.1	3.54	8.40	1.4	18.2	49	3.1	22.1	62	9.9	601	0
Sioux Falls, SD	22.6	42.3	-23	88	8.7	1.64	4.98	10.0	31.5	NA	6.1	17.3	64	12.5	1008	0
Spokane, WA	29.6	47.7	-10	74	11.3	1.49	3.79	4.1	16.4	53	4.3	18.9	55	9.5	815	0
St. Louis, MO	35.5	54.6	-5	92	11.6	3.58	9.52	4.4	28.8	53	6.5	16.5	61	11.9	617	0
Tampa, FL	56.5	76.6	29	91	6.8	3.01	12.64	T	T	71	10.2	10.8	55	9.6	84	128
Washington, DC	37.7	56.5	4	93	11.0	3.17	8.84	2.3	19.3	55	7.3	15.1	50	10.9	552	0
Wichita, KS	33.6	57.2	-3	92	7.8	2.43	9.17	2.7	16.8	61	8.8	15.1	52	14.2	608	0

INTERNATIONAL DATA

	AVG HIGH TEMP	AVG LOW TEMP	AVG PRECIP	DAYS WITH MEASURABLE PRECIP
Cape Town, South Africa	77	58	0.9	NA
Delhi, India	87	58	0.5	NA
Hong Kong, China	67	60	2.9	NA
Jerusalem, Israel	65	46	2.5	NA
London, United Kingdom	50	38	1.5	11
Manila, Philippines	92	71	0.3	3
Melbourne, Australia	75	55	2.1	9
Montreal, Canada	35	20	2.9	13
Moscow, Russia	34	24	1.3	9
Nairobi, Kenya	82	57	3.4	8
Paris, France	54	38	1.4	12
Rio de Janeiro, Brazil	83	72	5.1	12
Rome, Italy	59	45	2.2	8
Sydney, Australia	76	63	5.2	14
Tahiti, Society Islands	89	72	16.9	NA
Tokyo, Japan	54	36	4.2	10
Toronto, Canada	38	23	2.4	13

APRIL—U.S. CLIMATIC DATA

	TEMPERATURE (°F)					AMOUNT OF PRECIPITATION					MEAN			AVG	DEGREE DAYS (BASE 65°F)	
	AVERAGE		RECORD		MEAN DAYS OF PRECIP	TOTAL		SNOWFALL		% OF POSSIBLE SUN	CLEAR DAYS	CLOUDY DAYS	REL HUMID-ITY	WIND SPEED (MPH)	HEATING	COOLING
	MIN	MAX	MIN	MAX		AVG	MAX	AVG	MAX							
Albany, NY	35.1	57.5	9	93	12.2	2.99	7.95	3.1	17.7	53	5.8	16.1	55	10.6	558	0
Albuquerque, NM	39.6	70.8	18	89	3.3	0.52	4.20	0.5	8.1	77	12.8	7.8	18	11.1	301	7
Amarillo, TX	42.1	71.5	13	98	5.0	0.99	6.45	.05	8.3	73	11.6	9.6	38	15.5	266	20
Anchorage, AK	28.6	42.8	-21	72	7.1	0.67	1.91	5.5	27.6	52	5.3	18.3	54	7.2	876	0
Atlanta, GA	50.2	72.7	25	93	9.1	4.26	11.86	T	1.5	65	9.8	12.0	51	10.1	138	33
Atlantic City, NJ	39.3	60.7	12	94	10.7	3.56	7.95	0.3	6.0	55	7.4	13.2	64	11.9	450	0
Bakersfield, CA	50.1	75.9	30	101	4.2	0.57	2.39	0.0	—	NA	13.0	4.2	33	7.1	144	84
Bismarck, ND	31.0	54.9	-12	93	8.1	1.67	5.71	3.9	18.7	59	5.9	15.5	48	12.1	660	0
Boise, ID	36.7	61.4	11	92	8.2	1.24	4.73	0.7	8.0	66	6.7	14.7	36	10.2	477	0
Boston, MA	40.2	55.9	11	94	11.3	3.60	9.57	0.9	28.3	57	7.3	14.5	53	13.3	507	0
Buffalo, NY	36.2	54.2	5	94	14.3	2.87	5.90	3.3	15.7	52	4.9	16.9	58	12.8	594	0
Burlington, VT	34.2	53.6	2	91	12.4	2.76	6.55	3.9	21.3	49	5.1	17.3	58	9.3	633	0
Caribou, ME	29.0	46.7	-2	86	13.0	2.45	5.26	8.6	36.4	NA	5.0	18.0	65	11.7	813	0
Casper, WY	29.5	56.1	-6	84	10.4	1.56	5.75	13.6	56.3	NA	5.6	15.7	43	12.8	666	0
Charleston, SC	53.9	75.8	29	95	7.4	2.67	15.00	0.0	—	70	11.0	11.2	49	9.8	66	63
Charleston, WV	42.8	66.8	18	96	13.9	3.31	7.15	0.3	20.7	NA	5.8	16.5	49	7.8	312	6
Charlotte, NC	47.5	71.2	24	96	8.9	2.68	9.01	0.0	3.5	69	9.6	11.6	47	8.8	183	15
Chicago, IL	38.6	58.6	7	91	12.7	3.64	8.33	2.1	13.6	49	6.2	16.2	55	12.1	492	0
Cincinnati, OH	42.2	64.2	15	90	12.8	3.75	9.77	0.4	5.5	40	5.5	16.9	54	10.8	354	0
Cleveland, OH	37.3	57.9	10	88	14.3	3.14	6.61	2.2	14.5	53	5.2	16.9	56	11.8	522	0
Dallas, TX	54.7	76.3	29	101	8.2	3.50	13.04	0.0	0.1	64	8.6	13.5	53	12.7	75	90
Denver, CO	34.5	61.8	-2	90	8.8	1.71	8.24	9.3	33.8	68	6.8	12.4	35	10.2	504	0
Des Moines, IA	40.0	61.8	9	93	10.5	3.36	7.76	2.2	15.6	55	6.9	15.2	54	13.0	428	5
Detroit, MI	36.8	57.7	8	89	12.6	2.95	6.89	1.8	25.7	54	6.5	16.5	55	11.7	531	0
El Paso, TX	48.0	78.7	23	98	1.7	0.20	2.24	0.4	16.5	87	16.5	5.5	16	11.4	110	62
Ely, NV	26.0	57.0	-5	82	7.5	1.00	3.41	6.4	24.5	69	7.7	13.4	34	10.9	705	0
Grand Rapids, MI	35.4	56.6	3	90	13.2	3.37	8.29	3.3	15.6	51	6.5	16.6	57	11.0	570	0
Hartford, CT	37.5	59.9	9	96	11.0	3.85	9.90	1.8	14.3	56	6.5	14.8	52	10.1	489	0
Helena, MT	30.6	56.1	-10	86	8.1	0.97	3.00	4.9	21.7	58	3.8	17.5	38	9.3	648	0
Honolulu, HI	68.7	82.8	57	91	9.4	1.54	12.65	0.0	—	66	5.5	10.6	57	12.1	0	324
Houston, TX	58.1	78.4	31	95	7.0	3.21	10.92	0.0	—	51	7.3	16.2	59	9.4	36	135
Indianapolis, IN	41.5	63.3	16	90	12.3	3.70	8.60	0.5	6.9	54	5.9	16.7	58	11.3	378	0
Jackson, MS	51.9	77.4	30	94	8.8	5.57	15.95	T	T	63	9.1	13.6	59	8.5	87	75
Jacksonville, FL	54.9	79.1	34	95	6.6	2.77	11.61	0.0	—	71	10.1	9.5	65	8.8	37	97
Kansas City, MO	43.8	65.1	12	95	11.3	3.12	10.57	0.8	7.2	64	8.3	15.5	55	12.6	325	10
Las Vegas, NV	50.7	77.5	31	99	1.8	0.21	2.44	T	T	87	16.5	5.9	15	11.0	143	116
Lexington, KY	44.2	65.5	15	91	12.6	3.88	9.30	0.2	5.9	NA	6.2	15.4	55	11.0	312	6
Lincoln, NE	38.9	64.4	3	97	9.9	2.76	9.10	1.3	11.1	59	8.1	14.1	54	12.7	399	0
Little Rock, AR	52.2	72.9	28	95	10.3	5.08	14.81	T	T	62	8.7	13.8	56	9.2	130	58
Los Angeles, CA	54.2	72.3	36	106	3.5	1.03	7.53	0.0	—	70	11.6	9.4	60	8.5	128	77
Lubbock, TX	46.7	75.4	18	100	4.5	0.97	6.18	.02	5.3	74	12.8	8.9	32	14.9	161	44
Marquette, MI	27.0	47.5	-9	92	12.3	2.64	6.80	8.5	43.4	51	NA	NA	NA	NA	831	0
Medford, OR	38.0	64.6	21	96	9.2	1.16	4.40	0.2	4.2	NA	5.9	15.6	45	5.7	416	0
Memphis, TN	52.4	73.3	27	94	10.4	5.46	17.13	T	T	64	8.8	14.2	54	10.7	127	64
Miami, FL	67.8	82.4	46	96	6.0	2.85	17.29	0.0	—	79	8.3	6.9	55	10.6	0	306
Milwaukee, WI	35.8	52.9	12	82	12.0	3.50	7.31	1.9	15.8	53	6.5	15.6	62	13.0	618	0
Minneapolis, MN	36.2	56.5	2	95	10.2	2.42	7.00	3.2	21.8	56	6.8	15.3	52	12.3	558	0
Montgomery, AL	52.9	76.4	28	92	8.2	4.49	15.94	0.0	0.8	65	9.6	12.3	56	7.4	78	69
Nashville, TN	47.5	70.8	23	91	10.9	4.37	11.84	T	1.5	58	8.2	13.3	51	9.4	193	19
New Orleans, LA	58.4	78.5	32	92	7.1	4.50	16.12	0.0	—	63	7.8	11.4	60	9.5	28	133
New York, NY	43.8	61.2	12	96	10.6	4.20	11.51	0.9	13.5	59	7.6	11.9	51	10.5	375	0
Norfolk, VA	47.1	66.9	20	97	9.9	3.06	8.33	T	3.9	65	8.9	11.8	60	11.8	249	9
Oklahoma City, OK	48.8	71.9	20	100	7.8	2.77	11.91	T	4.1	66	9.5	12.7	49	14.5	176	38
Orlando, FL	59.4	83.0	38	96	8.5	1.80	9.10	0.0	—	NA	8.9	8.8	49	8.8	5	191
Philadelphia, PA	42.1	62.6	14	95	10.8	3.62	9.76	0.3	19.4	56	7.1	14.1	48	11.0	378	0
Phoenix, AZ	53.0	83.1	32	105	1.8	0.27	3.36	T	T	88	17.4	5.6	16	7.0	52	142
Pittsburgh, PA	38.8	60.3	11	90	13.4	3.15	8.11	1.4	13.5	48	4.5	17.5	50	10.5	462	0
Portland, OR	41.3	60.6	28	93	14.1	2.39	7.88	T	5.2	52	3.8	20.4	55	7.3	420	0
Providence, RI	37.7	57.0	11	98	10.9	4.11	12.74	0.8	18.0	57	7.9	13.8	58	12.2	528	0
Raleigh, NC	46.2	71.7	23	95	8.7	2.59	7.95	T	5.2	52	3.8	20.4	55	7.3	193	13
Reno, NV	33.3	63.7	13	89	4.1	0.38	3.00	1.4	12.7	80	8.8	11.1	28	8.2	492	0
Sacramento, CA	45.5	71.1	32	98	5.6	1.16	14.20	0.0	T	81	12.5	8.2	43	8.9	230	29
Salt Lake City, UT	37.9	61.3	14	86	9.6	2.12	4.90	5.3	26.4	67	7.0	13.8	39	9.5	464	0
San Antonio, TX	58.4	80.3	31	101	7.4	2.50	11.64	0.0	—	55	7.3	15.3	51	10.6	32	161
San Diego, CA	55.6	68.4	39	98	4.8	0.79	5.37	0.0	—	67	10.3	9.6	58	7.8	113	23
San Francisco, CA	49.8	62.1	31	92	6.0	1.29	10.06	0.0	—	73	11.2	9.6	60	12.1	275	0
Seattle, WA	41.2	57.2	29	87	13.9	2.33	6.53	0.1	2.4	53	2.8	19.9	57	9.6	474	0
Sioux Falls, SD	34.8	59.0	4	94	9.2	2.52	6.97	2.2	18.4	NA	6.9	14.9	53	13.3	543	0
Spokane, WA	34.7	57.0	14	90	8.5	1.18	3.97	0.7	6.6	60	4.5	17.2	44	9.9	573	0
St. Louis, MO	46.4	66.9	20	93	11.4	3.50	10.84	0.4	6.5	56	7.0	14.9	54	11.5	266	17
Tampa, FL	60.8	81.7	38	93	4.7	1.15	10.71	0.0	—	75	10.9	8.0	51	9.5	7	193
Washington, DC	46.4	66.7	15	95	9.8	2.71	9.13	0.0	5.0	58	7.1	14.0	47	10.5	264	9
Wichita, KS	44.5	68.3	15	98	8.0	2.38	12.42	0.3	6.5	63	8.7	13.3	50	14.2	278	20

INTERNATIONAL DATA

	AVG HIGH TEMP	AVG LOW TEMP	AVG PRECIP	DAYS WITH MEASURABLE PRECIP
Cape Town, South Africa	72	53	1.9	NA
Delhi, India	97	68	0.3	NA
Hong Kong, China	75	67	5.4	NA
Jerusalem, Israel	73	50	1.1	NA
London, United Kingdom	56	42	1.5	12
Manila, Philippines	94	74	0.8	3
Melbourne, Australia	68	51	2.3	11
Montreal, Canada	51	33	2.9	12
Moscow, Russia	49	36	1.4	8
Nairobi, Kenya	79	59	6.0	16
Paris, France	60	43	1.7	13
Rio de Janeiro, Brazil	80	69	4.2	10
Rome, Italy	66	50	2.0	6
Sydney, Australia	72	58	5.0	13
Tahiti, Society Islands	89	72	5.6	NA
Tokyo, Japan	63	46	5.3	11
Toronto, Canada	53	33	2.8	11

MAY—U.S. CLIMATIC DATA

	TEMPERATURE (°F)					AMOUNT OF PRECIPITATION					MEAN				DEGREE DAYS (BASE 65°F)	
	AVERAGE		RECORD		MEAN DAYS OF PRECIP	TOTAL		SNOWFALL		% OF POSSIBLE SUN	CLEAR DAYS	CLOUDY DAYS	REL HUMID-ITY	AVG WIND SPEED (MPH)	HEATING	COOLING
	MIN	MAX	MIN	MAX		AVG	MAX	AVG	MAX							
Albany, NY	45.4	69.7	26	97	13.1	3.41	8.96	0.1	5.4	55	5.2	16.6	60	9.0	247	18
Albuquerque, NM	48.6	79.7	28	98	4.2	0.50	3.56	0.0	2.0	80	14.8	6.1	18	10.6	89	64
Amarillo, TX	51.6	79.1	26	103	8.3	2.48	9.81	T	9.1	72	11.0	9.6	42	14.6	89	102
Anchorage, AK	38.8	54.4	1	82	6.7	0.73	2.00	0.5	8.8	52	3.9	20.4	49	8.3	570	0
Atlanta, GA	58.7	79.6	37	97	9.1	4.29	9.94	0.0	—	68	9.2	11.4	54	8.6	27	157
Atlantic City, NJ	49.6	71.2	25	99	10.4	3.33	11.51	T	T	54	6.4	14.0	69	10.4	167	25
Bakersfield, CA	57.3	84.6	34	110	1.6	0.20	2.99	0.0	—	NA	18.3	4.2	25	7.9	28	214
Bismarck, ND	42.2	67.8	13	102	9.8	2.18	7.04	0.9	10.3	62	6.2	14.3	44	11.8	324	14
Boise, ID	43.9	71.0	22	99	8.0	1.08	4.90	0.1	4.0	71	8.5	12.5	34	9.5	242	9
Boston, MA	49.8	66.6	31	97	11.7	3.25	13.38	0.0	0.5	59	6.5	14.5	59	12.1	221	10
Buffalo, NY	47.0	66.1	25	94	12.5	3.14	7.35	0.1	5.2	58	5.6	15.8	56	11.5	279	19
Burlington, VT	45.4	67.2	24	93	13.4	3.12	7.10	0.2	3.9	55	5.0	17.0	57	8.8	282	13
Caribou, ME	40.1	61.7	18	96	13.4	3.07	6.27	0.8	10.9	NA	4.0	18.0	61	11.4	437	0
Casper, WY	37.9	66.6	16	95	11.0	2.13	6.46	4.2	32.8	NA	5.1	15.1	41	11.6	394	0
Charleston, SC	62.9	82.7	36	99	9.2	4.01	9.56	0.0	—	71	7.9	12.2	54	8.7	0	242
Charleston, WV	51.5	75.5	26	98	13.2	3.94	8.76	T	0.2	NA	5.8	15.0	55	6.2	129	83
Charlotte, NC	56.4	78.3	32	100	9.7	3.82	12.48	0.0	—	69	8.0	13.1	53	7.6	42	116
Chicago, IL	47.7	70.1	24	98	11.0	3.32	7.59	0.1	2.2	56	7.0	14.0	54	10.5	235	46
Cincinnati, OH	51.8	74.0	27	95	11.5	4.28	9.52	T	1.5	51	5.9	15.6	55	8.8	151	86
Cleveland, OH	47.3	68.6	25	92	13.0	3.49	9.14	0.1	2.1	58	6.0	15.0	57	10.2	250	33
Dallas, TX	62.6	82.9	41	103	8.7	4.88	13.66	0.0	—	64	8.3	11.8	57	11.1	0	246
Denver, CO	43.6	70.8	19	96	10.7	2.40	8.57	1.9	15.5	64	6.0	12.8	38	9.4	253	11
Des Moines, IA	51.5	73.0	26	105	11.2	3.66	11.08	T	1.3	61	7.4	15.1	52	11.3	165	81
Detroit, MI	47.1	69.6	25	95	11.2	2.92	8.46	T	6.0	60	6.3	14.0	53	10.1	243	38
El Paso, TX	56.5	87.1	31	105	2.1	0.25	4.22	0.0	—	89	18.8	4.3	16	10.6	7	217
Ely, NV	33.7	67.3	7	89	7.1	1.15	3.26	2.5	12.1	72	7.7	11.9	31	10.7	450	0
Grand Rapids, MI	45.6	69.3	21	98	10.8	3.13	10.01	T	5.5	54	6.5	15.1	53	9.6	273	40
Hartford, CT	47.6	71.6	28	97	11.5	4.12	12.00	T	1.3	58	5.3	15.8	56	8.9	194	27
Helena, MT	39.6	65.4	17	95	11.1	1.78	6.67	1.5	13.9	50	4.7	16.5	38	8.9	388	
Honolulu, HI	70.3	84.7	60	92	7.3	1.13	7.23	0.0	—	68	6.5	94	54	12.1	0	388
Houston, TX	64.4	84.6	44	99	8.4	5.24	16.88	0.0	—	57	7.0	13.4	50	8.3	0	295
Indianapolis, IN	51.7	73.8	28	96	12.3	4.00	10.10	T	2.4	60	7.1	14.9	57	9.5	165	96
Jackson, MS	60.0	84.0	38	100	9.3	5.05	12.23	0.0	—	63	8.7	11.6	60	7.2	7	224
Jacksonville, FL	62.1	84.7	45	100	8.3	3.55	14.80	0.0	—	69	9.3	9.8	66	8.2	0	260
Kansas City, MO	53.9	74.3	27	103	11.4	5.04	12.75	T	1.7	64	7.5	14.6	58	10.2	135	107
Las Vegas, NV	60.2	87.8	38	109	1.4	0.28	0.96	0.0	—	88	18.6	4.7	13	10.9	14	293
Lexington, KY	53.5	74.3	26	96	11.9	4.47	10.91	0.0	6.0	66	7.2	13.6	56	9.0	135	104
Lincoln, NE	50.0	74.2	25	104	11.2	3.90	11.33	0.1	3.0	62	8.0	13.0	55	10.5	161	72
Little Rock, AR	59.7	79.6	39	98	10.0	5.24	15.91	0.0	—	68	8.0	12.3	59	7.8	31	177
Los Angeles, CA	57.7	73.9	40	103	1.2	0.19	3.57	0.0	—	66	10.1	10.2	66	8.3	72	97
Lubbock, TX	55.8	83.1	29	109	7.4	2.35	7.80	0.0	0.0	73	11.3	8.4	36	14.1	25	162
Marquette, MI	38.4	61.9	17	100	10.4	3.03	8.09	1.5	22.6	57	NA	NA	NA	NA	471	13
Medford, OR	43.4	72.9	28	103	8.1	1.00	4.58	T	T	NA	9.3	12.8	39	5.7	219	8
Memphis, TN	61.2	81.0	38	99	9.2	4.98	13.34	0.0	—	69	8.7	12.6	55	8.9	25	217
Miami, FL	72.1	85.3	50	96	10.4	6.21	18.66	0.0	—	70	5.9	10.0	60	9.5	0	425
Milwaukee, WI	44.8	64.3	21	103	12.0	2.84	9.05	0.0	3.2	59	7.0	14.0	60	11.7	338	16
Minneapolis, MN	47.6	69.4	18	106	11.3	3.39	10.33	0.2	3.0	60	7.0	14.8	50	11.2	244	43
Montgomery, AL	60.8	82.9	40	99	8.5	3.92	12.01	0.0	—	65	9.2	11.6	60	6.2	6	220
Nashville, TN	56.6	78.8	34	97	10.7	4.88	11.04	0.0	T	60	8.3	12.9	56	7.6	59	143
New Orleans, LA	65.2	84.4	41	97	7.6	4.56	21.18	0.0	—	61	9.6	10.2	60	8.1	0	304
New York, NY	53.7	71.7	32	99	11.1	4.42	9.74	T	T	61	8.0	10.7	53	8.8	125	54
Norfolk, VA	56.8	75.3	32	100	9.9	3.81	10.12	0.0	T	65	7.9	13.4	66	10.4	51	85
Oklahoma City, OK	57.7	79.1	32	104	10.0	5.22	12.07	0.0	—	67	8.9	12.2	54	12.7	31	136
Orlando, FL	65.9	87.8	48	102	5.4	3.55	10.36	0.0	—	NA	10.2	8.3	46	9.4	0	369
Philadelphia, PA	52.7	73.1	28	97	11.4	3.75	9.46	T	T	56	6.1	14.4	53	9.6	123	58
Phoenix, AZ	61.5	92.4	39	114	0.9	0.14	1.31	0.0	—	93	21.3	3.5	13	7.0	0	376
Pittsburgh, PA	48.4	70.6	26	95	12.4	3.59	6.56	0.2	3.1	52	5.2	16.6	51	9.0	214	44
Portland, OR	47.0	67.1	29	100	11.7	2.06	6.60	0.0	0.6	58	4.8	18.9	53	7.0	249	0
Providence, RI	47.3	67.3	29	95	11.4	3.76	9.25	0.2	7.0	57	6.7	14.3	63	10.9	246	7
Raleigh, NC	55.3	78.6	31	99	10.3	3.92	9.90	0.0	T	60	8.2	12.7	54	7.7	47	109
Reno, NV	40.1	72.9	16	98	4.3	0.69	2.89	1.0	14.1	82	12.1	9.1	25	7.9	274	10
Sacramento, CA	50.3	80.3	36	107	2.7	0.27	3.25	0.0	—	88	17.8	5.2	35	9.3	80	89
Salt Lake City, UT	45.6	71.9	25	99	8.0	1.80	5.76	0.7	7.5	72	9.2	11.4	33	9.4	215	23
San Antonio, TX	65.7	85.3	43	104	8.3	4.22	14.07	0.0	—	56	6.2	13.6	55	10.2	0	326
San Diego, CA	59.1	69.1	45	98	2.3	0.19	2.54	0.0	—	58	8.6	11.1	64	7.8	73	45
San Francisco, CA	50.5	62.7	36	106	2.7	0.25	4.02	0.0	—	72	13.9	7.5	60	13.3	260	0
Seattle, WA	46.3	63.9	28	93	10.2	1.70	4.76	T	T	56	4.4	17.5	54	9.0	307	0
Sioux Falls, SD	45.9	70.7	17	104	10.4	3.03	9.42	T	3.0	NA	7.3	13.8	49	11.9	240	35
Spokane, WA	41.9	65.8	24	97	9.2	1.41	3.97	0.1	3.5	63	5.6	15.0	41	9.0	344	0
St. Louis, MO	56.0	76.1	31	96	10.8	3.97	12.92	T	4.0	61	7.6	13.8	56	9.5	111	145
Tampa, FL	67.5	87.2	49	98	6.4	3.10	17.64	0.0	—	75	10.2	8.7	53	8.9	0	378
Washington, DC	56.6	76.2	28	99	11.2	3.66	10.69	T	T	59	7.3	13.8	52	9.2	60	104
Wichita, KS	54.3	76.9	27	100	10.6	3.81	11.22	0.0	—	65	8.3	12.7	54	12.5	102	121

INTERNATIONAL DATA

	AVG HIGH TEMP	AVG LOW TEMP	AVG PRECIP	DAYS WITH MEASURABLE PRECIP
Cape Town, South Africa	67	49	3.7	NA
Delhi, India	105	79	0.5	NA
Hong Kong, China	82	74	11.5	NA
Jerusalem, Israel	81	57	0.1	NA
London, United Kingdom	62	47	1.8	12
Manila, Philippines	94	76	4.0	8
Melbourne, Australia	62	47	2.2	14
Montreal, Canada	65	45	2.6	12
Moscow, Russia	63	48	2.1	6
Nairobi, Kenya	76	58	5.0	14
Paris, France	67	49	2.2	12
Rio de Janeiro, Brazil	77	66	3.1	10
Rome, Italy	74	56	1.8	5
Sydney, Australia	67	52	4.8	13
Tahiti, Society Islands	87	70	4.0	NA
Tokyo, Japan	71	54	5.8	12
Toronto, Canada	65	43	2.6	11

JUNE—U.S. CLIMATIC DATA

	TEMPERATURE (°F)					AMOUNT OF PRECIPITATION					MEAN				DEGREE DAYS (BASE 65°F)	
	AVERAGE		RECORD		MEAN DAYS OF PRECIP	TOTAL		SNOWFALL		% OF POSSIBLE SUN	CLEAR DAYS	CLOUDY DAYS	REL HUMID-ITY	AVG WIND SPEED (MPH)		
	MIN	MAX	MIN	MAX		AVG	MAX	AVG	MAX						HEATING	COOLING
Albany, NY	54.6	79.0	35	100	11.2	3.62	8.74	0.0	—	59	5.3	13.6	64	8.2	34	91
Albuquerque, NM	58.3	90.0	40	107	3.7	0.59	8.15	0.0	—	83	17.9	3.5	17	10.0	0	279
Amarillo, TX	60.7	87.6	38	108	8.3	3.70	10.73	0.0	0.3	77	13.0	5.8	44	14.3	6	279
Anchorage, AK	47.2	61.6	29	86	8.4	1.14	3.40	0.0	0.4	48	2.5	20.5	55	8.2	318	0
Atlanta, GA	66.2	85.8	39	102	9.9	3.56	11.21	0.0	—	67	7.6	10.2	57	8.0	0	330
Atlantic City, NJ	58.7	80.0	37	106	8.9	2.64	8.45	0.0	—	58	6.9	12.1	71	9.3	12	144
Bakersfield, CA	64.0	92.4	38	114	0.5	0.10	1.11	0.0	—	NA	23.5	1.9	23	7.9	6	402
Bismarck, ND	51.6	77.1	30	111	11.8	2.72	9.90	0.0	2.0	64	7.4	12.2	48	10.5	116	98
Boise, ID	52.1	80.9	30	109	6.4	0.81	3.41	T	T	75	11.5	8.3	30	9.1	75	120
Boston, MA	59.1	76.3	41	100	10.5	3.09	13.20	0.0	—	64	6.9	12.6	59	11.3	32	113
Buffalo, NY	56.5	75.3	35	97	10.3	3.55	9.67	T	T	65	6.3	12.2	57	11.0	59	86
Burlington, VT	54.6	75.8	33	100	12.5	3.47	9.92	0.0	—	58	4.8	14.6	62	8.3	58	64
Caribou, ME	49.1	71.9	30	96	13.6	2.91	7.11	T	T	NA	3.3	16.9	65	10.4	143	8
Casper, WY	46.9	78.6	28	102	8.5	1.46	4.71	0.2	4.5	NA	9.6	9.6	33	11.0	139	70
Charleston, SC	69.1	87.6	49	104	10.8	6.43	27.24	0.0	—	68	6.2	12.5	59	8.3	0	399
Charleston, WV	59.8	83.1	33	105	11.3	3.59	10.67	0.0	—	NA	4.7	12.2	61	5.6	10	202
Charlotte, NC	65.6	85.8	45	103	9.8	3.39	11.04	0.0	—	70	7.4	11.5	57	6.9	0	32
Chicago, IL	57.5	79.6	35	104	10.3	3.78	10.58	0.0	T	67	7.3	11.3	55	9.1	35	143
Cincinnati, OH	60.0	82.0	39	102	10.5	3.84	9.86	0.0	—	73	6.9	12.8	56	7.9	11	191
Cleveland, OH	56.8	78.3	31	104	11.0	3.70	10.73	0.0	—	65	6.7	12.0	58	9.4	40	118
Dallas, TX	70.0	91.9	51	113	5.9	2.98	12.18	0.0	—	71	11.3	7.3	50	10.8	0	480
Denver, CO	52.4	81.4	30	104	8.9	1.79	4.96	0.0	0.4	71	9.6	8.1	35	8.9	71	128
Des Moines, IA	61.2	82.2	37	103	10.8	4.46	15.71	0.0	—	68	8.1	11.4	53	10.3	10	214
Detroit, MI	56.3	78.9	36	104	10.5	3.61	8.31	0.0	—	66	7.8	11.1	54	9.0	38	116
El Paso, TX	64.3	96.5	46	114	3.3	0.67	3.18	0.0	—	89	20.0	2.8	18	9.5	0	462
Ely, NV	40.7	78.3	18	99	4.8	0.88	3.53	.02	5.6	80	13.3	6.5	23	10.6	188	26
Grand Rapids, MI	55.3	78.7	32	102	10.2	3.68	13.22	0.0	—	61	6.4	12.3	56	8.8	50	110
Hartford, CT	56.9	80.0	37	100	11.3	3.75	13.60	0.0	—	60	5.5	14.0	60	8.1	20	125
Helena, MT	48.3	75.8	30	102	11.7	1.87	5.63	0.1	2.7	62	5.6	13.4	38	8.6	137	50
Honolulu, HI	72.2	86.5	64	92	6.0	0.50	4.26	0.0	—	70	5.5	7.1	53	12.8	0	432
Houston, TX	70.6	90.1	52	103	7.9	4.96	19.21	0.0	—	64	9.1	8.8	59	7.7	0	462
Indianapolis, IN	37.0	82.7	39	102	10.0	3.49	12.21	0.0	—	66	7.0	12.3	58	8.5	5	212
Jackson, MS	67.1	90.6	47	105	8.0	3.18	9.69	0.0	—	70	9.7	8.4	60	6.3	0	414
Jacksonville, FL	69.1	89.3	47	103	12.0	5.69	23.32	0.0	—	63	5.7	11.2	73	8.0	0	423
Kansas City, MO	63.1	83.3	42	108	10.2	4.72	11.86	0.0	—	69	9.6	10.8	58	9.9	7	253
Las Vegas, NV	69.4	100.3	48	116	0.7	0.12	0.97	0.0	—	92	22.4	2.5	10	11.0	0	597
Lexington, KY	61.5	82.7	39	104	10.7	3.66	11.69	0.0	—	NA	7.4	11.3	57	8.1	5	221
Lincoln, NE	60.2	84.7	39	109	7.9	3.89	12.93	0.0	—	70	9.7	10.4	51	10.1	11	236
Little Rock, AR	67.5	87.4	46	105	8.1	3.26	9.28	0.0	—	73	9.5	8.9	57	7.4	0	375
Los Angeles, CA	61.1	78.3	46	112	0.5	0.03	1.39	0.0	—	65	9.3	9.6	67	7.9	35	176
Lubbock, TX	64.3	90.0	49	114	7.0	2.76	8.48	0.0	0.0	77	13.5	5.6	37	11.2	0	366
Marquette, MI	47.4	70.9	25	101	12.3	3.48	6.61	T	0.2	58	NA	NA	NA	NA	193	19
Medford, OR	50.7	82.1	31	111	5.3	0.58	3.49	0.0	T	NA	12.6	9.2	33	5.9	60	105
Memphis, TN	68.9	89.3	48	104	8.4	3.57	18.16	0.0	—	74	10.1	8.8	56	8.0	0	423
Miami, FL	75.1	87.6	60	98	15.0	9.33	25.34	0.0	—	75	3.3	12.5	66	8.2	0	492
Milwaukee, WI	55.0	74.9	33	104	11.1	3.24	10.03	0.0	—	64	7.7	12.1	61	10.5	82	82
Minneapolis, MN	57.6	78.8	34	104	12.0	4.05	9.00	0.0	—	64	7.5	12.4	53	10.5	41	137
Montgomery, AL	67.9	89.4	48	106	9.0	3.90	15.59	0.0	—	65	8.6	9.9	59	5.9	0	411
Nashville, TN	64.7	86.5	42	106	9.4	3.57	11.95	0.0	—	66	8.3	9.4	55	7.0	0	318
New Orleans, LA	70.8	89.2	50	102	10.2	5.84	17.62	0.0	—	67	9.4	7.9	62	6.8	0	450
New York, NY	63.0	80.1	44	101	10.1	3.67	10.27	0.0	—	64	8.0	9.6	55	8.1	0	203
Norfolk, VA	65.2	82.9	40	104	9.0	3.82	10.53	0.0	—	68	7.6	10.8	67	9.6	0	277
Oklahoma City, OK	66.1	87.3	46	107	8.5	4.31	14.76	0.0	—	74	10.8	8.6	54	12.3	0	351
Orlando, FL	71.8	90.5	53	100	13.8	7.32	18.28	0.0	—	NA	4.5	11.4	57	8.0	0	483
Philadelphia, PA	61.8	81.7	44	102	10.3	3.74	10.06	0.0	—	62	7.0	11.8	54	8.7	5	209
Phoenix, AZ	70.6	102.3	49	122	0.7	0.17	1.70	0.0	—	94	23.3	2.2	12	6.9	0	645
Pittsburgh, PA	56.9	78.9	34	98	11.3	3.71	10.29	0.0	—	57	5.3	13.2	52	8.0	36	123
Portland, OR	52.9	74.0	39	102	9.6	1.48	5.38	T	T	55	5.9	16.7	49	7.1	91	46
Providence, RI	56.8	76.9	39	101	10.8	3.33	11.08	0.0	—	60	6.9	12.8	67	10.0	31	88
Raleigh, NC	63.6	85.0	38	104	9.2	3.68	10.44	0.0	—	61	7.6	10.5	56	7.0	0	279
Reno, NV	46.9	83.1	25	103	3.1	0.46	1.94	T	0.2	85	16.8	5.6	22	7.5	76	79
Sacramento, CA	55.3	87.8	41	115	1.1	0.12	1.45	0.0	—	93	21.7	2.5	31	9.8	12	210
Salt Lake City, UT	55.4	82.8	32	104	5.6	0.93	3.84	T	2.0	79	13.6	6.3	26	9.4	51	174
San Antonio, TX	72.6	91.8	48	107	5.9	3.81	11.95	0.0	—	67	7.0	7.6	51	10.2	0	516
San Diego, CA	61.9	71.6	50	101	1.0	0.07	0.87	0.0	—	57	9.1	9.1	66	7.7	51	105
San Francisco, CA	52.6	64.1	41	106	1.1	0.15	2.57	0.0	—	73	16.1	5.3	59	13.9	198	0
Seattle, WA	51.9	69.9	38	100	9.3	1.50	3.90	0.0	—	54	4.9	17.4	53	8.8	144	21
Sioux Falls, SD	56.1	80.5	32	110	10.9	3.40	8.43	0.0	—	NA	8.6	10.3	51	10.7	50	149
Spokane, WA	49.2	74.7	33	101	7.8	1.26	5.12	T	T	66	7.2	12.4	36	9.0	139	49
St. Louis, MO	65.7	85.2	43	104	9.4	3.72	12.35	0.0	—	66	7.4	11.7	56	8.8	0	312
Tampa, FL	72.9	89.5	53	99	11.6	5.48	18.52	0.0	—	67	5.4	10.6	50	8.2	0	480
Washington, DC	66.5	84.7	36	102	9.4	3.38	18.19	0.0	—	65	7.9	11.3	53	8.7	0	318
Wichita, KS	64.6	86.8	43	110	9.3	4.31	14.43	0.0	—	69	9.6	9.9	49	12.2	5	326

INTERNATIONAL DATA

	AVG HIGH TEMP	AVG LOW TEMP	AVG PRECIP	DAYS WITH MEASURABLE PRECIP
Cape Town, South Africa	65	46	4.3	NA
Delhi, India	102	83	2.9	NA
Hong Kong, China	85	78	15.5	NA
Jerusalem, Israel	85	60	0	NA
London, United Kingdom	69	53	1.8	11
Manila, Philippines	91	75	10.1	16
Melbourne, Australia	57	44	2.0	14
Montreal, Canada	75	55	3.2	12
Moscow, Russia	69	53	2.6	7
Nairobi, Kenya	75	54	1.3	5
Paris, France	73	55	2.1	12
Rio de Janeiro, Brazil	76	64	2.1	7
Rome, Italy	82	63	1.5	4
Sydney, Australia	62	48	5.2	12
Tahiti, Society Islands	86	69	3.0	NA
Tokyo, Japan	76	63	6.5	12
Toronto, Canada	75	53	2.6	9

JULY—U.S. CLIMATIC DATA

	TEMPERATURE (°F)				MEAN	AMOUNT OF PRECIPITATION				% OF	MEAN		REL	AVG WIND	DEGREE DAYS (BASE 65°F)	
	AVERAGE		RECORD		DAYS OF	TOTAL		SNOWFALL		POSSIBLE	CLEAR	CLOUDY	HUMID-	SPEED		
	MIN	MAX	MIN	MAX	PRECIP	AVG	MAX	AVG	MAX	SUN	DAYS	DAYS	ITY	(MPH)	HEATING	COOLING
Albany, NY	59.6	84.0	40	104	10.2	3.18	9.37	0.0	—	64	6.1	12.0	63	7.5	0	213
Albuquerque, NM	64.4	92.5	54	105	8.8	1.37	4.45	0.0	—	76	12.1	4.7	27	9.1	0	419
Amarillo, TX	65.5	91.7	51	106	8.2	2.62	10.73	0.0	—	78	13.1	5.6	41	12.6	0	422
Anchorage, AK	51.7	65.2	34	84	11.6	1.71	4.57	0.0	—	43	3.4	22.0	61	7.1	205	0
Atlanta, GA	69.5	88.0	53	105	12.0	5.01	17.71	0.0	—	63	5.6	12.0	61	7.5	0	428
Atlantic City, NJ	64.8	84.5	42	104	8.6	3.83	15.69	0.0	—	60	7.3	13.0	70	8.6	0	301
Bakersfield, CA	69.6	98.5	46	118	0.1	0.01	0.43	0.0	—	NA	26.6	1.3	21	7.2	0	592
Bismarck, ND	56.4	84.4	32	114	8.9	2.14	13.75	0.0	1.5	75	11.5	6.7	42	9.2	15	183
Boise, ID	57.7	90.2	35	111	2.3	0.35	2.01	T	T	87	20.9	3.1	22	8.4	6	285
Boston, MA	65.1	81.8	50	104	9.1	2.84	11.69	0.0	—	67	7.0	11.6	55	10.9	0	264
Buffalo, NY	61.9	80.2	43	97	9.8	3.08	8.93	0.0	—	68	7.0	10.9	54	10.4	5	194
Burlington, VT	59.7	81.2	39	100	11.9	3.65	9.31	0.0	—	64	5.3	12.7	60	7.9	6	176
Caribou, ME	54.5	76.5	36	95	14.0	4.01	6.83	0.0	—	NA	3.0	15.1	70	9.8	61	76
Casper, WY	54.0	87.6	30	104	7.8	1.26	3.05	0.0	—	NA	14.1	5.9	26	10.1	10	190
Charleston, SC	72.7	90.2	58	104	13.7	6.84	18.46	0.0	—	68	4.9	13.8	63	7.9	0	512
Charleston, WV	64.4	85.7	46	108	12.9	4.99	13.54	0.0	—	NA	4.2	13.5	66	5.1	0	313
Charlotte, NC	69.6	88.9	53	103	11.3	3.92	16.55	0.0	—	68	6.5	13.0	58	6.6	0	443
Chicago, IL	62.6	83.7	40	106	9.8	3.66	9.56	0.0	—	69	8.8	9.2	57	8.1	5	259
Cincinnati, OH	64.8	85.5	47	109	10.0	4.24	10.02	0.0	—	70	7.4	11.9	57	7.1	0	313
Cleveland, OH	61.4	82.4	41	103	10.1	3.52	9.15	0.0	T	67	8.6	10.5	57	8.7	0	218
Dallas, TX	74.1	96.5	56	110	4.9	2.31	11.13	0.0	—	81	15.3	6.4	44	9.5	0	629
Denver, CO	58.6	88.2	42	105	9.1	1.91	6.99	0.0	—	72	9.2	6.0	34	8.4	0	267
Des Moines, IA	66.5	86.7	47	110	9.1	3.78	10.51	0.0	—	72	10.5	9.2	55	8.9	0	360
Detroit, MI	61.3	83.3	41	105	9.1	3.18	8.76	0.0	—	70	9.3	9.5	53	8.3	0	231
El Paso, TX	68.4	96.1	56	112	7.5	1.54	6.50	0.0	—	80	12.3	5.7	28	8.4	0	536
Ely, NV	48.0	87.0	28	101	5.0	0.69	2.30	0.0	—	80	14.9	4.9	22	10.3	14	92
Grand Rapids, MI	60.4	82.8	41	108	9.1	3.19	9.35	0.0	—	64	8.5	10.3	56	8.1	0	208
Hartford, CT	62.2	85.0	44	102	9.6	3.19	11.24	0.0	—	64	5.9	13.0	61	7.5	0	270
Helena, MT	53.4	85.0	36	105	7.4	1.10	4.70	T	T	78	14.8	5.3	29	7.8	34	164
Honolulu, HI	73.5	87.5	66	94	7.5	0.59	2.56	0.0	—	73	7.5	5.4	52	13.5	0	481
Houston, TX	72.4	92.7	55	104	9.7	3.60	14.80	0.0	—	66	7.7	8.1	58	6.8	0	546
Indianapolis, IN	65.2	85.5	44	106	9.2	4.47	13.12	0.0	—	67	8.5	9.9	62	7.4	0	322
Jackson, MS	70.5	92.4	51	107	10.5	4.51	13.25	0.0	—	65	7.5	10.1	66	5.9	0	512
Jacksonville, FL	71.9	91.4	61	105	14.7	5.60	16.21	0.0	—	61	4.7	12.0	75	7.2	0	515
Kansas City, MO	68.2	88.7	51	112	7.1	4.38	15.47	0.0	—	74	14.5	7.3	53	8.9	0	419
Las Vegas, NV	76.2	105.9	56	117	2.6	0.35	2.48	0.0	—	87	19.6	3.4	15	10.1	0	809
Lexington, KY	65.7	85.8	47	108	11.2	5.00	11.24	0.0	—	NA	7.8	10.5	59	7.4	0	335
Lincoln, NE	66.3	90.0	42	115	8.1	3.20	11.40	0.0	—	73	12.3	8.3	49	9.8	0	409
Little Rock, AR	71.3	91.5	54	112	8.3	3.31	9.23	0.0	—	71	8.8	9.3	60	6.8	0	508
Los Angeles, CA	64.5	84.0	49	109	0.5	0.01	0.32	0.0	—	82	12.7	5.4	67	7.7	0	293
Lubbock, TX	68.0	91.9	43	109	6.8	2.37	7.20	0.0	0.0	79	14.2	5.9	39	11.2	0	465
Marquette, MI	53.5	76.6	34	108	9.8	2.88	10.20	0.0	0.0	64	NA	NA	NA	NA	74	78
Medford, OR	55.2	90.5	38	115	1.5	0.26	2.74	0.0	—	NA	23.3	2.6	26	5.8	8	253
Memphis, TN	72.9	92.3	52	108	8.7	3.79	9.96	0.0	—	74	10.4	8.8	57	7.5	0	546
Miami, FL	76.2	89.0	66	98	15.8	5.70	15.22	0.0	—	79	2.5	11.3	63	7.8	0	546
Milwaukee, WI	62.0	79.9	40	105	9.5	3.47	7.66	0.0	—	70	10.3	9.5	61	9.6	14	197
Minneapolis, MN	63.1	84.0	43	108	9.7	3.53	17.90	0.0	—	71	10.4	9.1	53	9.3	11	278
Montgomery, AL	71.5	91.1	59	107	11.9	5.19	13.42	0.0	—	62	5.4	11.5	66	5.7	0	505
Nashville, TN	68.9	89.5	51	107	10.2	3.97	9.43	0.0	—	64	8.3	9.8	57	6.5	0	443
New Orleans, LA	73.1	90.6	52	102	14.8	6.12	13.15	0.0	—	62	5.1	11.5	66	6.1	0	524
New York, NY	68.4	85.2	52	107	10.4	4.35	12.33	0.0	—	65	8.5	9.5	55	7.6	0	366
Norfolk, VA	70.0	86.4	50	106	11.3	5.06	14.37	0.0	—	64	7.5	11.8	70	8.9	0	409
Oklahoma City, OK	70.6	93.4	53	110	6.4	2.61	11.90	0.0	—	79	14.5	7.1	45	10.9	0	527
Orlando, FL	73.1	91.5	64	100	17.4	7.25	19.57	0.0	—	NA	3.2	11.1	59	7.4	0	536
Philadelphia, PA	67.2	86.1	51	104	9.1	4.28	10.42	0.0	—	62	7.1	12.0	54	8.0	0	363
Phoenix, AZ	79.5	105.0	61	121	4.3	0.74	6.47	0.0	—	85	16.3	4.4	20	7.1	0	846
Pittsburgh, PA	61.6	82.6	42	103	10.8	3.75	9.51	0.0	—	59	5.3	12.5	53	7.3	6	227
Portland, OR	56.5	79.9	43	107	3.8	0.63	2.68	0.0	—	70	13.0	9.3	45	7.6	28	127
Providence, RI	63.2	82.1	48	102	8.7	3.18	8.08	0.0	—	64	7.2	11.8	66	9.5	0	239
Raleigh, NC	68.1	88.0	48	105	11.2	4.01	12.36	0.0	—	61	7.4	11.9	59	6.6	0	406
Reno, NV	51.3	91.9	33	108	2.4	0.28	1.75	0.0	T	92	22.3	2.6	19	6.9	13	217
Sacramento, CA	58.1	93.2	47	114	0.3	0.05	1.89	0.0	—	97	27.1	0.9	28	9.0	0	332
Salt Lake City, UT	63.7	92.2	40	107	4.5	0.81	2.57	0.0	—	83	16.9	4.3	21	9.5	0	400
San Antonio, TX	75.0	95.0	60	106	4.3	2.16	16.92	0.0	—	74	9.2	6.6	45	9.3	0	620
San Diego, CA	65.7	76.2	54	100	0.3	0.02	0.92	0.0	—	69	13.3	4.8	65	7.3	13	199
San Francisco, CA	53.5	64.6	43	104	0.3	0.04	0.62	0.0	—	66	20.6	2.9	59	13.5	187	0
Seattle, WA	55.2	75.2	43	100	5.1	0.76	2.39	T	T	65	10.4	10.4	49	8.3	58	64
Sioux Falls, SD	62.3	86.3	34	110	9.3	2.68	9.11	0.0	—	NA	11.7	7.3	50	9.7	10	298
Spokane, WA	54.4	83.1	37	108	4.1	0.67	2.37	0.0	—	80	16.8	5.8	26	8.3	30	148
St. Louis, MO	70.4	89.3	51	112	8.4	3.85	10.71	0.0	—	70	9.5	9.8	56	7.9	0	459
Tampa, FL	74.5	90.2	63	97	15.7	6.58	20.59	0.0	—	62	2.2	12.8	63	7.4	0	530
Washington, DC	71.4	88.5	45	106	9.7	3.80	11.06	0.0	—	64	7.8	11.4	53	8.2	0	465
Wichita, KS	69.9	92.8	51	113	7.5	3.13	13.37	0.0	—	76	12.8	7.6	44	11.1	0	508

INTERNATIONAL DATA

	AVG HIGH TEMP	AVG LOW TEMP	AVG PRECIP	DAYS WITH MEASURABLE PRECIP
Cape Town, South Africa	63	46	3.7	NA
Delhi, India	96	81	7.1	NA
Hong Kong, China	87	78	15.0	NA
Jerusalem, Israel	87	63	0.0	NA
London, United Kingdom	71	56	2.2	12
Manila, Philippines	89	75	11.2	20
Melbourne, Australia	56	42	1.9	15
Montreal, Canada	79	60	3.5	12
Moscow, Russia	71	57	2.9	8
Nairobi, Kenya	73	53	0.5	4
Paris, France	76	58	2.3	12
Rio de Janeiro, Brazil	75	63	1.6	7
Rome, Italy	87	67	0.6	1
Sydney, Australia	60	46	4.1	11
Tahiti, Society Islands	86	68	2.1	NA
Tokyo, Japan	83	70	5.6	11
Toronto, Canada	70	60	2.9	10

AUGUST—U.S. CLIMATIC DATA

	TEMPERATURE (°F)					AMOUNT OF PRECIPITATION					MEAN				DEGREE DAYS (BASE 65°F)	
	AVERAGE		RECORD		MEAN DAYS OF PRECIP	TOTAL		SNOWFALL		% OF POSSIBLE SUN	CLEAR DAYS	CLOUDY DAYS	REL HUMID-ITY	AVG WIND SPEED (MPH)		
	MIN	MAX	MIN	MAX		AVG	MAX	AVG	MAX						HEATING	COOLING
Albany, NY	57.8	81.4	34	102	10.1	3.47	10.59	0.0	—	60	6.9	12.5	70	7.0	12	155
Albuquerque, NM	50.0	89.0	52	101	9.3	1.64	4.90	0.0	—	76	13.7	4.8	30	8.2	0	338
Amarillo, TX	63.8	89.1	48	106	8.4	3.22	7.55	0.0	—	77	14.6	6.3	46	12.0	0	357
Anchorage, AK	49.5	63.0	31	85	12.6	2.44	5.91	0.0	—	41	3.5	21.5	64	6.6	270	0
Atlanta, GA	69.0	87.1	55	102	9.4	3.66	10.02	0.0	—	65	7.5	10.4	61	7.2	0	406
Atlantic City, NJ	63.5	83.3	40	104	8.6	4.14	16.12	0.0	—	63	7.4	12.1	74	8.2	0	260
Bakersfield, CA	68.5	96.6	44	117	0.4	0.09	1.18	0.0	—	NA	26.1	1.4	24	6.8	0	546
Bismarck, ND	53.9	82.7	32	109	8.4	1.72	6.55	0.0	0.2	73	11.8	8.1	39	9.5	69	171
Boise, ID	34.0	88.1	37	112	2.7	0.43	2.37	0.0	—	84	18.5	4.6	23	8.2	20	252
Boston, MA	64.0	79.8	46	102	9.9	3.24	17.09	0.0	—	66	9.2	11.0	58	10.7	6	220
Buffalo, NY	60.1	77.9	38	99	10.7	4.17	10.67	0.0	—	64	6.8	12.3	59	9.8	17	141
Burlington, VT	57.9	77.9	35	101	12.4	4.06	11.54	0.0	—	60	6.0	13.3	67	7.4	29	119
Caribou, ME	52.1	73.6	34	95	13.0	4.07	12.09	0.0	—	NA	4.3	15.0	73	9.3	115	47
Casper, WY	51.8	85.7	33	102	5.5	0.67	2.79	T	T	NA	13.4	7.2	25	10.4	36	151
Charleston, SC	72.2	89.0	56	105	12.3	7.22	19.18	0.0	—	66	5.9	11.9	63	7.4	0	484
Charleston, WV	63.4	84.4	41	108	11.0	4.01	10.45	0.0	—	NA	4.8	11.7	70	4.4	0	276
Charlotte, NC	68.9	87.7	53	103	9.4	3.73	14.61	0.0	—	69	7.8	10.5	58	6.4	0	412
Chicago, IL	61.6	81.8	42	102	9.1	4.22	17.10	0.0	—	67	9.1	10.5	57	8.1	19	226
Cincinnati, OH	62.9	84.1	43	103	9.1	3.35	11.72	0.0	—	71	8.1	10.8	58	6.7	0	266
Cleveland, OH	60.3	80.5	38	102	9.6	3.40	8.96	0.0	—	63	8.7	11.2	60	8.4	11	178
Dallas, TX	73.6	96.2	56	110	4.7	2.21	10.82	0.0	—	77	15.0	6.0	45	9.0	0	617
Denver, CO	56.9	85.8	40	105	8.7	1.51	5.85	0.0	—	72	9.9	7.3	35	8.1	0	203
Des Moines, IA	63.6	84.2	40	110	9.2	4.20	13.68	0.0	—	70	10.8	9.8	58	8.7	11	287
Detroit, MI	59.6	81.3	38	104	9.2	3.43	8.33	0.0	—	68	8.7	10.9	56	8.2	16	186
El Paso, TX	66.6	93.5	52	108	7.4	1.58	6.85	0.0	—	81	14.0	4.7	32	7.9	0	468
Ely, NV	46.5	84.4	24	97	5.3	0.83	2.51	0.0	—	81	14.8	4.5	23	10.4	62	77
Grand Rapids, MI	58.4	80.5	39	102	9.3	3.57	8.46	0.0	—	61	8.5	11.6	58	7.8	18	157
Hartford, CT	60.4	82.7	36	102	9.9	3.65	21.87	0.0	—	63	6.3	13.6	65	7.1	6	210
Helena, MT	28.0	83.2	29	105	7.5	1.29	4.23	0.0	6.2	74	13.3	7.0	30	7.5	65	139
Honolulu, HI	74.2	88.7	66	96	6.5	0.44	3.93	0.0	—	75	8.0	6.4	53	13.2	0	508
Houston, TX	72.0	92.5	54	108	9.8	3.49	15.43	0.0	—	64	6.6	8.5	59	6.0	0	536
Indianapolis, IN	62.8	83.6	41	103	8.7	3.64	8.34	0.0	—	69	9.1	10.0	66	7.1	6	260
Jackson, MS	69.7	92.0	54	107	9.7	3.77	11.39	0.0	—	64	8.9	9.1	68	5.7	0	493
Jacksonville, FL	71.8	90.7	63	102	14.4	7.93	16.24	0.0	—	60	5.1	10.7	78	6.9	0	502
Kansas City, MO	65.7	86.4	46	113	8.6	4.01	11.64	0.0	—	68	12.8	8.1	57	9.0	6	350
Las Vegas, NV	74.2	103.2	54	116	3.1	0.49	2.59	0.0	—	88	21.4	2.9	18	9.4	0	735
Lexington, KY	64.4	84.9	42	105	9.2	3.93	11.18	0.0	—	NA	9.5	9.4	58	7.0	0	301
Lincoln, NE	63.3	86.7	42	110	8.6	3.41	14.21	0.0	—	72	11.9	8.9	54	9.6	9	319
Little Rock, AR	70.0	89.9	52	111	7.1	3.10	14.46	0.0	—	73	11.6	8.5	60	6.5	0	465
Los Angeles, CA	65.7	84.5	49	106	0.4	0.14	2.47	0.0	—	83	13.3	5.5	68	7.6	0	316
Lubbock, TX	66.2	89.6	43	107	6.7	2.51	8.85	0.0	0.0	78	15.3	6.0	41	9.9	0	400
Marquette, MI	51.7	73.3	31	102	13.1	3.41	8.59	T	T	57	NA	NA	NA	NA	122	45
Medford, OR	55.1	89.9	39	114	2.1	0.52	2.83	0.0	—	NA	21.5	3.7	28	5.3	11	244
Memphis, TN	71.1	90.8	48	107	8.0	3.43	10.60	0.0	—	75	11.8	7.7	57	7.0	0	496
Miami, FL	76.7	89.0	67	98	17.1	7.58	16.88	0.0	—	75	2.5	10.7	65	7.8	0	552
Milwaukee, WI	60.8	77.8	42	101	9.0	3.53	8.06	0.0	—	66	10.4	10.0	63	9.4	27	160
Minneapolis, MN	60.3	80.7	39	103	9.9	3.62	9.31	0.0	—	68	10.4	9.5	56	9.2	22	192
Montgomery, AL	56.0	90.4	57	104	9.1	3.69	15.58	0.0	—	65	8.1	9.1	68	5.2	0	484
Nashville, TN	67.7	88.4	47	105	8.9	3.46	9.60	0.0	—	64	10.2	9.0	59	6.1	0	406
New Orleans, LA	72.8	90.2	60	103	13.3	6.17	22.74	0.0	—	60	7.5	9.6	66	5.9	0	512
New York, NY	67.3	83.7	46	104	9.8	4.01	17.41	0.0	—	64	9.2	9.7	57	7.6	0	326
Norfolk, VA	69.4	85.1	47	105	10.3	4.81	14.87	0.0	—	65	8.3	10.9	74	8.8	0	378
Oklahoma City, OK	69.6	92.5	49	113	6.4	2.60	8.34	0.0	—	79	14.9	6.5	45	10.6	0	499
Orlando, FL	73.4	91.5	64	100	15.9	6.78	16.11	0.0	—	NA	3.1	10.9	60	7.2	0	543
Philadelphia, PA	66.3	84.6	44	106	9.1	3.80	12.10	0.0	—	62	8.3	11.7	54	7.8	0	326
Phoenix, AZ	77.5	102.3	58	116	4.8	1.02	5.56	0.0	—	85	17.6	3.8	23	6.6	0	77
Pittsburgh, PA	60.2	80.8	39	103	9.9	3.21	7.86	0.0	—	57	6.3	12.9	56	6.9	14	184
Portland, OR	56.9	80.3	43	107	5.1	1.09	4.53	0.0	—	65	11.1	10.5	46	7.1	35	147
Providence, RI	61.9	80.7	40	104	9.5	3.63	12.24	0.0	—	60	8.3	12.3	70	9.3	8	203
Raleigh, NC	67.5	86.8	46	105	9.9	4.02	13.63	0.0	—	61	7.6	11.1	60	6.4	0	375
Reno, NV	49.6	89.6	24	105	2.3	0.32	1.65	0.0	T	93	22.1	2.6	20	6.4	20	162
Sacramento, CA	58.0	92.1	48	111	0.4	0.07	0.65	0.0	—	96	25.6	1.5	29	8.6	0	313
Salt Lake City, UT	61.8	89.4	37	106	5.7	0.86	3.66	0.0	—	82	15.6	4.6	23	9.6	0	329
San Antonio, TX	74.5	95.3	57	108	5.3	2.54	11.14	0.0	—	73	10.1	6.1	46	8.6	0	617
San Diego, CA	67.3	77.8	54	98	0.5	0.10	2.13	0.0	—	69	14.9	4.5	66	7.2	0	240
San Francisco, CA	54.6	65.6	42	100	0.5	0.07	0.78	0.0	—	65	19.2	3.2	61	12.8	156	0
Seattle, WA	55.7	75.2	44	99	6.6	1.14	5.49	0.0	—	64	8.8	12.6	52	7.9	65	81
Sioux Falls, SD	59.4	83.3	34	109	8.9	2.85	9.33	0.0	—	NA	12.0	8.6	52	9.8	22	220
Spokane, WA	54.3	82.5	35	104	5.1	0.72	2.12	0.0	—	77	15.2	7.4	28	8.1	56	161
St. Louis, MO	67.9	87.3	47	108	7.8	2.85	20.45	0.0	—	65	10.4	9.6	58	7.7	0	391
Tampa, FL	74.5	90.2	66	98	16.7	7.61	18.59	0.0	—	60	2.8	11.8	65	7.1	0	530
Washington, DC	70.0	86.9	38	106	9.2	3.91	14.41	0.0	—	64	9.3	11.8	55	8.0	0	419
Wichita, KS	67.9	90.7	45	114	7.1	3.02	11.96	0.0	—	75	13.7	7.5	45	11.1	0	443

INTERNATIONAL DATA

	AVG HIGH TEMP	AVG LOW TEMP	AVG PRECIP	DAYS WITH MEASURABLE PRECIP
Cape Town, South Africa	64	46	3.3	NA
Delhi, India	93	79	6.8	NA
Hong Kong, China	87	78	14.2	NA
Jerusalem, Israel	87	64	0.0	NA
London, United Kingdom	71	56	2.3	11
Manila, Philippines	88	75	17.0	23
Melbourne, Australia	59	44	2.0	15
Montreal, Canada	77	58	3.6	12
Moscow, Russia	69	54	2.9	7
Nairobi, Kenya	73	53	0.7	5
Paris, France	75	58	2.5	13
Rio de Janeiro, Brazil	76	64	1.7	7
Rome, Italy	86	67	0.8	2
Sydney, Australia	63	48	3.2	11
Tahiti, Society Islands	86	68	1.7	NA
Tokyo, Japan	86	72	6.0	10
Toronto, Canada	77	58	2.7	9

SEPTEMBER—U.S. CLIMATIC DATA

	TEMPERATURE (°F)					AMOUNT OF PRECIPITATION						MEAN			AVG	DEGREE DAYS (BASE 65°F)	
	AVERAGE		RECORD		MEAN DAYS OF PRECIP	TOTAL		SNOWFALL		% OF POSSIBLE SUN		CLEAR DAYS	CLOUDY DAYS	REL HUMID-ITY	AVG WIND SPEED (MPH)		
	MIN	MAX	MIN	MAX		AVG	MAX	AVG	MAX							HEATING	COOLING

City	AVERAGE MIN	AVERAGE MAX	RECORD MIN	RECORD MAX	MEAN DAYS OF PRECIP	TOTAL AVG	TOTAL MAX	SNOWFALL AVG	SNOWFALL MAX	% OF POSSIBLE SUN	MEAN CLEAR DAYS	MEAN CLOUDY DAYS	REL HUMIDITY	AVG WIND SPEED (MPH)	HEATING	COOLING
Albany, NY	49.4	73.2	24	100	9.5	2.95	9.90	0.0	—	57	7.8	12.2	75	7.4	141	30
Albuquerque, NM	55.2	81.9	35	100	5.7	1.00	3.31	T	T	79	16.9	5.3	31	8.6	18	126
Amarillo, TX	56.4	81.8	30	103	5.9	1.99	6.42	T	0.3	73	15.2	7.6	49	13.0	26	149
Anchorage, AK	41.6	55.2	19	73	13.8	2.70	7.35	0.2	6.3	41	4.0	20.7	63	6.2	498	0
Atlanta, GA	63.5	81.8	36	102	7.5	3.42	14.26	0.0	—	64	9.7	10.5	60	8.0	10	241
Atlantic City, NJ	55.5	76.6	32	99	7.4	2.93	14.73	0.0	—	60	10.0	11.7	78	8.6	53	86
Bakersfield, CA	63.5	90.1	30	112	0.9	0.17	1.15	0.0	—	NA	23.8	1.9	29	6.2	8	362
Bismarck, ND	43.1	70.8	10	105	7.0	1.49	6.93	0.3	5.0	66	10.4	10.8	44	10.0	262	22
Boise, ID	48.2	77.0	23	102	3.7	0.80	2.93	0.0	T	81	16.9	6.0	30	8.3	160	88
Boston, MA	56.8	72.8	34	102	8.7	3.06	10.94	0.0	—	64	10.5	11.3	60	11.1	72	66
Buffalo, NY	53.0	70.8	32	98	10.8	3.49	8.99	T	T	59	6.5	13.6	60	10.4	130	37
Burlington, VT	48.8	69.0	25	98	11.6	3.30	10.25	T	T	54	6.0	14.0	74	8.1	199	16
Caribou, ME	43.2	64.0	23	91	12.3	3.45	8.14	T	T	NA	5.5	15.3	75	10.4	343	0
Casper, WY	41.6	73.8	16	97	6.3	0.94	3.40	1.2	11.5	NA	13.4	8.0	30	11.0	250	34
Charleston, SC	67.9	84.9	42	100	9.4	4.73	17.31	0.0	—	63	6.6	12.8	62	7.9	0	342
Charleston, WV	56.5	78.8	33	104	9.2	3.24	7.69	0.0	—	NA	6.7	12.0	71	4.8	44	125
Charlotte, NC	62.9	81.9	38	104	7.2	3.50	12.66	0.0	—	67	9.2	11.4	57	6.7	6	228
Chicago, IL	53.9	74.8	28	101	9.6	3.82	11.44	T	T	58	9.0	11.3	58	8.8	84	66
Cincinnati, OH	56.6	77.9	32	101	8.0	2.88	8.61	0.0	—	64	9.5	11.6	58	7.4	51	120
Cleveland, OH	54.2	73.6	32	101	9.8	3.44	11.05	T	T	60	8.4	12.2	60	9.1	99	66
Dallas, TX	66.9	87.8	36	111	6.8	3.39	10.01	0.0	—	74	13.1	8.4	54	9.4	0	372
Denver, CO	47.6	76.9	17	97	6.1	1.24	4.67	1.6	21.3	75	13.5	7.5	34	8.0	144	63
Des Moines, IA	54.5	75.6	26	102	8.5	3.53	14.81	T	0.3	65	12.0	10.8	60	9.5	71	74
Detroit, MI	52.5	73.9	29	100	9.7	2.89	7.52	0.0	—	61	8.3	12.2	57	8.7	102	48
El Paso, TX	61.6	87.1	41	104	5.2	1.70	6.68	0.0	—	82	17.7	5.2	34	7.8	0	282
Ely, NV	37.3	75.2	15	93	4.5	1.01	4.99	0.3	6.3	82	17.1	5.0	24	10.3	274	13
Grand Rapids, MI	49.9	72.0	27	98	10.4	4.24	11.85	T	T	54	7.0	14.0	61	8.3	139	19
Hartford, CT	51.8	74.8	30	101	9.4	3.79	14.59	0.0	—	60	8.3	12.5	70	7.2	96	45
Helena, MT	41.0	69.8	6	99	6.6	1.15	3.66	1.6	14.1	67	10.2	11.1	35	7.4	321	33
Honolulu, HI	73.5	88.5	66	95	7.0	0.78	7.74	0.0	—	75	8.2	5.9	52	11.6	0	480
Houston, TX	67.9	88.4	45	109	9.5	4.89	13.37	0.0	—	62	8.9	10.6	62	6.8	0	396
Indianapolis, IN	55.6	77.6	28	100	7.7	2.87	10.37	0.0	—	66	10.8	10.6	66	7.9	58	106
Jackson, MS	63.7	88.0	35	107	8.4	3.55	14.77	0.0	—	61	10.4	10.5	70	6.5	6	333
Jacksonville, FL	69.0	87.2	48	100	13.1	7.05	21.79	0.0	—	55	5.3	12.9	80	7.8	0	393
Kansas City, MO	56.9	78.1	31	109	7.6	4.86	16.17	0.0	T	65	11.8	10.0	59	9.3	56	131
Las Vegas, NV	66.2	94.7	43	113	1.8	0.28	3.39	0.0	—	91	22.7	2.3	17	8.8	0	465
Lexington, KY	58.0	78.3	32	103	8.1	3.20	9.69	0.0	—	NA	10.6	10.9	59	7.7	47	143
Lincoln, NE	53.2	77.2	26	106	7.8	3.48	8.32	T	0.8	66	12.4	10.3	57	9.8	81	90
Little Rock, AR	63.7	82.9	37	106	7.2	4.08	10.23	0.0	—	68	11.2	10.1	64	6.8	11	260
Los Angeles, CA	64.6	82.7	44	110	1.0	0.45	5.67	0.0	—	79	12.9	6.6	65	7.2	10	271
Lubbock, TX	59.4	82.9	33	105	5.8	2.60	8.85	0.0	0.0	70	14.5	7.7	45	10.5	17	200
Marquette, MI	44.1	63.8	21	98	14.8	4.08	12.73	0.2	6.0	47	NA	NA	NA	NA	330	0
Medford, OR	48.2	82.8	29	107	4.2	0.86	4.22	0.0	—	NA	17.7	5.7	31	4.5	93	108
Memphis, TN	64.5	83.9	36	103	7.0	3.53	12.34	0.0	—	70	12.4	10.0	56	7.5	10	286
Miami, FL	75.9	87.8	67	97	17.6	7.63	24.40	0.0	—	72	2.3	12.9	67	8.2	0	507
Milwaukee, WI	52.8	70.6	28	99	9.0	3.38	9.87	T	T	59	9.7	11.1	63	10.5	123	24
Minneapolis, MN	50.3	70.7	26	104	9.4	2.72	7.77	0.0	1.7	61	10.1	11.3	59	9.9	167	32
Montgomery, AL	66.1	87.0	39	106	7.5	4.09	12.00	0.0	—	63	10.0	11.0	67	6.0	0	345
Nashville, TN	61.1	82.5	39	105	7.8	3.46	11.44	0.0	—	63	10.8	10.4	59	6.4	21	225
New Orleans, LA	69.5	86.6	42	101	9.8	5.51	18.98	0.0	—	63	9.9	9.8	65	7.3	0	393
New York, NY	60.1	76.2	39	102	8.3	3.89	16.85	0.0	—	63	10.6	9.4	57	8.1	34	130
Norfolk, VA	64.2	79.6	35	102	7.8	3.90	13.80	0.0	—	64	9.3	11.2	75	9.6	11	218
Oklahoma City, OK	62.2	83.8	35	108	6.7	3.84	11.86	0.0	—	72	13.6	8.4	52	11.2	15	255
Orlando, FL	72.4	89.7	56	97	13.8	6.01	15.87	0.0	—	NA	3.8	11.8	60	7.7	0	480
Philadelphia, PA	58.7	77.6	35	102	7.9	3.42	13.07	0.0	—	60	9.8	11.3	55	8.2	32	128
Phoenix, AZ	70.9	98.2	47	118	3.0	0.64	5.41	0.0	—	89	22.0	3.0	23	6.3	0	588
Pittsburgh, PA	53.5	74.3	31	102	9.3	2.97	10.08	0.0	—	58	7.6	12.1	56	7.4	100	67
Portland, OR	52.0	74.6	34	102	7.8	1.75	5.52	T	T	61	10.1	11.9	49	6.5	102	51
Providence, RI	53.8	74.3	33	100	8.2	3.48	9.74	0.0	—	61	9.7	12.1	72	9.4	90	63
Raleigh, NC	61.1	81.1	37	104	7.4	3.19	21.79	0.0	—	59	10.0	11.3	59	6.8	9	192
Reno, NV	41.3	79.5	20	101	2.5	0.39	2.41	T	1.5	92	20.7	3.7	23	5.7	178	40
Sacramento, CA	55.7	87.3	43	108	1.2	0.37	3.62	0.0	—	93	23.7	2.2	31	7.7	16	211
Salt Lake City, UT	51.0	79.2	27	100	5.3	1.28	7.04	0.1	4.0	83	16.9	5.1	29	9.1	108	114
San Antonio, TX	69.2	89.3	41	112	7.1	3.41	15.78	0.0	—	67	9.6	8.3	52	8.6	0	429
San Diego, CA	65.6	77.1	50	111	1.1	0.24	2.58	0.0	—	68	15.0	5.8	65	6.9	19	211
San Francisco, CA	55.9	68.7	38	103	0.9	0.26	5.07	0.0	—	72	18.4	3.6	57	11.0	116	35
Seattle, WA	51.9	69.3	35	98	9.4	1.88	5.95	T	T	59	7.6	13.6	58	8.1	156	24
Sioux Falls, SD	48.7	73.1	13	104	8.1	3.02	9.26	T	0.9	NA	11.7	10.4	56	10.3	165	42
Spokane, WA	45.8	72.0	22	98	5.8	0.73	5.58	0.0	1.4	70	11.9	9.8	35	8.2	223	40
St. Louis, MO	60.5	79.9	36	104	7.9	3.12	10.53	0.0	—	64	11.7	10.2	61	8.1	21	177
Tampa, FL	72.8	89.0	54	96	13.1	5.98	18.93	0.0	—	61	4.5	12.0	62	8.0	0	480
Washington, DC	62.5	80.1	30	104	7.6	3.31	17.45	0.0	—	62	10.1	11.7	55	8.2	14	203
Wichita, KS	59.2	81.4	31	108	7.7	3.49	10.69	T	T	68	13.1	10.2	51	11.6	29	188

INTERNATIONAL DATA

	AVG HIGH TEMP	AVG LOW TEMP	AVG PRECIP	DAYS WITH MEASURABLE PRECIP
Cape Town, South Africa	65	49	2.3	NA
Delhi, India	93	75	4.6	NA
Hong Kong, China	85	77	10.1	NA
Jerusalem, Israel	85	62	0.0	NA
London, United Kingdom	65	52	2.0	13
Manila, Philippines	88	74	12.7	22
Melbourne, Australia	63	46	2.3	14
Montreal, Canada	68	49	3.5	12
Moscow, Russia	58	45	2.3	8
Nairobi, Kenya	78	54	0.8	4
Paris, France	70	53	2.2	13
Rio de Janeiro, Brazil	75	65	2.6	11
Rome, Italy	79	62	2.5	5
Sydney, Australia	67	51	2.7	11
Tahiti, Society Islands	86	69	2.1	NA
Tokyo, Japan	79	66	9.2	13
Toronto, Canada	70	49	2.5	9

OCTOBER—U.S. CLIMATIC DATA

	TEMPERATURE (°F)				MEAN DAYS OF PRECIP	AMOUNT OF PRECIPITATION				% OF POSSIBLE SUN	MEAN		REL HUMID-ITY	AVG WIND SPEED (MPH)	DEGREE DAYS (BASE 65°F)	
	AVERAGE		RECORD			TOTAL		SNOWFALL								
	MIN	MAX	MIN	MAX		AVG	MAX	AVG	MAX		CLEAR DAYS	CLOUDY DAYS			HEATING	COOLING
Albany, NY	38.6	61.8	16	91	8.7	2.83	13.48	0.1	6.5	51	7.7	14.0	72	8.0	459	0
Albuquerque, NM	43.0	71.0	21	91	4.8	0.89	5.40	0.0	2.5	79	17.6	5.9	29	8.3	259	11
Amarillo, TX	44.5	72.5	12	99	5.0	1.37	7.64	T	4.4	74	16.1	8.0	44	13.0	226	25
Anchorage, AK	28.7	40.5	-6	63	12.3	2.03	4.11	7.4	28.1	37	5.0	21.2	65	6.5	942	0
Atlanta, GA	51.9	72.7	28	95	6.3	3.05	11.06	0.0	T	68	14.1	9.6	54	8.4	138	54
Atlantic City, NJ	43.7	66.0	23	91	7.2	2.82	12.13	T	—	56	10.2	12.3	78	9.1	324	10
Bakersfield, CA	54.8	80.7	31	104	1.7	0.29	2.04	0.0	—	NA	19.5	4.9	33	5.5	60	147
Bismarck, ND	32.5	58.7	-10	95	5.7	0.90	4.30	1.3	23.7	58	9.5	13.8	51	10.1	598	0
Boise, ID	39.0	64.6	11	94	6.2	0.75	4.06	0.1	2.7	68	11.9	8.4	40	8.4	414	0
Boston, MA	46.9	62.7	25	90	8.9	3.30	10.66	0.0	1.1	60	10.8	12.2	58	11.9	321	5
Buffalo, NY	42.7	59.4	20	92	11.4	3.09	9.13	0.2	22.6	51	6.6	16.2	60	11.2	431	0
Burlington, VT	38.6	57.0	15	85	11.4	2.88	6.75	0.2	5.1	48	6.1	17.0	71	8.6	533	0
Caribou, ME	34.4	52.0	14	79	12.3	3.10	8.73	2.0	12.1	NA	4.8	18.3	75	10.9	676	0
Casper, WY	32.2	60.5	-3	87	6.6	0.97	2.66	5.0	22.1	NA	11.4	11.7	41	12.2	577	0
Charleston, SC	56.3	77.2	27	95	5.9	2.90	14.32	0.0	—	66	11.7	10.9	55	8.1	74	130
Charleston, WV	44.2	68.2	17	96	9.7	2.89	7.18	0.2	2.8	NA	8.7	13.1	66	5.2	299	26
Charlotte, NC	50.6	72.0	24	98	6.8	3.36	14.72	0.0	T	68	13.3	10.2	54	7.0	161	47
Chicago, IL	42.2	63.3	14	91	9.1	2.41	12.06	0.3	6.6	49	8.5	13.2	56	9.9	391	12
Cincinnati, OH	44.2	66.0	16	92	8.2	2.86	9.51	0.0	6.2	37	10.7	13.2	56	8.1	327	20
Cleveland, OH	43.5	62.1	22	90	11.0	2.54	9.50	0.7	8.0	53	8.3	14.9	60	10.0	387	8
Dallas, TX	55.8	78.5	26	102	6.1	3.52	14.18	0.0	—	61	13.7	9.8	55	9.6	51	119
Denver, CO	36.4	66.3	-2	90	5.4	0.98	4.17	3.7	31.2	72	13.4	8.5	35	8.0	429	7
Des Moines, IA	42.7	64.3	7	95	7.6	2.62	7.29	0.3	7.4	61	11.6	12.3	58	10.4	372	15
Detroit, MI	40.9	61.5	17	91	9.1	2.10	7.80	0.1	3.8	50	7.7	14.4	57	9.6	435	7
El Paso, TX	49.6	78.4	25	96	4.2	0.76	5.15	0.0	1.0	83	19.0	5.4	30	7.7	88	57
Ely, NV	28.2	63.5	-3	85	5.0	0.89	3.67	2.4	12.1	75	14.4	8.4	32	10.1	595	0
Grand Rapids, MI	39.1	59.8	17	89	11.0	2.81	8.32	0.5	8.4	43	5.6	17.2	62	9.3	485	0
Hartford, CT	40.7	63.7	17	91	8.3	3.57	16.32	0.1	1.7	57	9.0	13.1	66	7.7	397	0
Helena, MT	31.6	58.5	-8	87	5.8	0.60	2.78	2.1	27.7	60	8.2	14.2	42	7.1	617	0
Honolulu, HI	72.3	86.9	63	94	8.9	2.28	11.26	0.0	—	68	7.5	8.4	55	10.8	0	453
Houston, TX	57.6	81.6	32	99	7.7	4.27	17.64	0.0	—	61	11.2	11.1	57	6.7	31	174
Indianapolis, IN	43.5	65.8	17	90	8.0	2.63	8.98	T	9.3	61	11.1	12.4	64	8.8	338	18
Jackson, MS	50.3	79.1	26	98	6.2	3.26	10.58	0.0	—	66	14.8	9.9	72	6.6	104	94
Jacksonville, FL	59.3	80.2	36	96	8.7	2.90	16.25	0.0	—	57	9.8	11.9	79	8.2	31	179
Kansas City, MO	45.7	67.5	17	98	8.0	3.29	11.94	0.0	6.5	60	12.5	10.9	57	10.4	279	18
Las Vegas, NV	54.3	82.1	26	103	1.8	0.21	1.45	T	T	87	20.6	4.1	19	8.1	62	164
Lexington, KY	46.0	67.2	20	93	7.9	2.57	7.95	0.0	2.8	NA	12.1	11.7	57	8.4	287	30
Lincoln, NE	40.5	66.7	3	99	7.1	2.12	6.03	0.4	13.2	63	11.3	12.2	57	10.1	362	8
Little Rock, AR	52.6	73.1	27	97	6.7	3.62	15.35	T	0.2	69	14.4	9.5	63	6.9	126	60
Los Angeles, CA	60.3	79.0	40	108	1.9	0.31	6.96	0.0	—	73	13.1	7.9	59	6.9	17	162
Lubbock, TX	48.1	74.7	18	100	5.2	1.86	13.93	0.2	7.5	71	16.4	8.0	46	11.1	154	42
Marquette, MI	34.8	52.3	9	89	15.8	3.61	7.59	3.8	18.6	40	NA	NA	NA	NA	663	0
Medford, OR	40.4	69.4	18	99	7.9	1.49	9.16	T	1.3	NA	10.0	12.6	46	3.7	302	7
Memphis, TN	51.9	74.3	25	95	6.1	3.01	10.13	0.0	T	70	14.3	9.6	51	7.8	131	72
Miami, FL	72.1	84.5	51	95	14.5	5.64	25.02	0.0	—	72	6.4	10.9	64	9.3	0	412
Milwaukee, WI	41.8	58.7	15	89	8.7	2.41	7.15	0.1	4.0	55	9.5	12.9	63	11.4	456	0
Minneapolis, MN	38.8	58.8	10	90	8.1	2.19	6.42	0.4	8.2	55	10.3	13.5	59	10.5	502	0
Montgomery, AL	53.3	78.2	26	100	5.7	2.45	9.06	0.0	T	65	13.8	9.7	66	5.7	91	116
Nashville, TN	48.3	72.5	26	94	7.0	2.62	8.34	0.0	1.0	62	12.6	10.1	55	6.7	195	52
New Orleans, LA	58.7	79.4	35	94	5.5	3.05	25.11	0.0	—	67	14.9	8.2	59	7.5	30	157
New York, NY	49.7	65.3	25	94	8.3	3.56	16.73	0.0	1.2	61	11.8	9.5	55	8.9	250	17
Norfolk, VA	52.9	69.5	23	98	7.8	3.15	10.12	0.0	2.9	59	11.4	12.6	75	10.4	164	46
Oklahoma City, OK	50.4	73.6	16	97	6.4	3.23	13.18	T	1.3	69	14.3	9.8	54	12.0	137	44
Orlando, FL	65.8	84.6	43	95	8.5	2.42	14.51	0.0	—	NA	9.4	10.0	56	8.6	0	316
Philadelphia, PA	46.4	66.3	25	96	7.6	2.62	8.68	0.1	2.2	59	10.5	12.1	54	8.8	283	17
Phoenix, AZ	59.1	87.7	34	107	2.7	0.63	4.40	0.0	—	88	20.5	4.4	22	5.9	13	273
Pittsburgh, PA	42.3	62.5	16	91	10.5	2.36	8.20	0.2	8.4	52	8.0	14.5	54	8.4	400	9
Portland, OR	44.9	64.0	26	92	12.8	2.67	11.63	0.0	0.6	42	5.4	18.7	63	6.5	326	0
Providence, RI	43.0	64.1	20	90	8.3	3.69	15.38	0.2	2.5	60	11.0	12.1	70	9.6	359	6
Raleigh, NC	48.4	71.6	19	98	7.1	2.86	10.23	0.0	T	60	12.7	11.3	54	7.1	189	37
Reno, NV	32.9	68.6	8	91	3.0	0.38	2.14	0.4	6.8	83	15.7	7.7	28	5.3	440	0
Sacramento, CA	50.4	77.9	34	104	3.4	1.08	7.51	0.0	—	85	19.2	5.8	39	6.6	78	53
Salt Lake City, UT	40.2	66.1	16	89	6.4	1.44	4.76	1.4	27.8	72	14.1	9.0	41	8.5	373	7
San Antonio, TX	58.8	81.7	27	99	6.5	3.17	18.07	0.0	—	63	11.7	9.6	53	8.6	30	191
San Diego, CA	60.9	74.6	43	107	2.3	0.37	4.98	0.0	—	68	14.1	7.4	60	6.5	24	107
San Francisco, CA	55.2	68.7	34	102	3.8	1.26	7.30	0.0	—	70	15.6	6.5	59	9.3	123	30
Seattle, WA	45.8	59.7	28	89	13.5	3.23	8.95	0.1	2.0	44	4.0	19.6	67	8.6	378	0
Sioux Falls, SD	36.0	61.2	9	94	6.2	1.78	6.89	0.5	10.0	NA	11.2	12.3	59	10.8	508	0
Spokane, WA	36.0	58.6	9	87	7.9	0.99	5.41	0.5	6.1	52	7.7	15.6	49	8.1	549	0
St. Louis, MO	48.3	68.5	21	94	8.4	2.68	8.74	T	0.5	59	12.2	11.1	61	8.8	237	33
Tampa, FL	65.2	84.3	40	95	6.8	2.02	10.33	0.0	—	65	11.2	9.1	57	8.7	0	307
Washington, DC	50.3	69.1	15	96	7.4	3.02	9.41	0.0	2.2	59	11.0	12.5	53	8.6	195	31
Wichita, KS	46.6	70.6	14	96	6.3	2.22	9.42	T	2.5	65	13.0	10.8	55	12.0	221	22

INTERNATIONAL DATA

	AVG HIGH TEMP	AVG LOW TEMP	AVG PRECIP	DAYS WITH MEASURABLE PRECIP
Cape Town, South Africa	70	52	1.6	NA
Delhi, India	93	65	0.4	NA
Hong Kong, China	81	73	4.5	NA
Jerusalem, Israel	81	59	0.5	NA
London, United Kingdom	58	46	2.2	13
Manila, Philippines	88	73	7.4	17
Melbourne, Australia	67	49	2.7	14
Montreal, Canada	56	39	3.0	13
Moscow, Russia	45	35	2.0	10
Nairobi, Kenya	80	56	1.9	7
Paris, France	60	46	2.0	13
Rio de Janeiro, Brazil	77	66	3.1	13
Rome, Italy	71	55	3.9	8
Sydney, Australia	71	56	3.0	12
Tahiti, Society Islands	87	70	3.5	NA
Tokyo, Japan	69	55	8.2	12
Toronto, Canada	58	39	2.4	11

NOVEMBER—U.S. CLIMATIC DATA

	Temperature (°F) Average		Record		Mean Days of Precip	Amount of Precipitation Total		Snowfall		% of Possible Sun	Mean		Rel Humid-ity	Avg Wind Speed (MPH)	Degree Days (Base 65°F)	
	Min	Max	Min	Max		Avg	Max	Avg	Max		Clear Days	Cloudy Days			Heating	Cooling
Albany, NY	30.7	48.7	-11	82	11.9	3.23	8.07	4.2	24.6	36	3.7	18.5	73	9.0	759	0
Albuquerque, NM	31.2	57.3	-7	77	3.3	0.43	4.70	1.1	9.3	77	15.2	7.1	36	7.8	621	0
Amarillo, TX	32.3	59.7	0	87	3.3	0.69	5.09	1.7	13.9	71	14.5	8.8	48	13.1	570	0
Anchorage, AK	15.1	27.2	-21	62	9.6	1.11	2.84	10.2	32.4	35	5.4	19.7	73	6.1	1314	0
Atlanta, GA	42.8	63.4	3	84	8.3	3.86	15.72	0.0	2.2	60	11.8	12.2	54	9.1	367	0
Atlantic City, NJ	35.8	55.7	10	84	9.3	3.58	9.65	0.4	16.7	48	7.7	13.6	75	10.6	576	0
Bakersfield, CA	44.7	66.8	22	95	3.6	0.70	3.04	0.0	—	NA	12.0	10.0	49	5.1	283	7
Bismarck, ND	17.8	39.3	-30	79	5.9	0.49	3.10	5.1	31.0	45	6.2	16.8	63	10.0	1092	0
Boise, ID	31.1	48.7	-7	78	10.0	1.48	4.43	1.9	18.6	44	6.2	17.6	60	8.5	753	0
Boston, MA	38.3	52.2	-2	83	10.9	4.22	11.03	1.1	17.8	51	8.0	14.7	60	12.8	591	0
Buffalo, NY	33.9	47.1	2	80	15.8	3.83	6.37	12.4	45.6	29	2.1	22.4	69	12.8	735	0
Burlington, VT	29.6	44.0	-3	75	14.1	3.13	10.13	6.9	30.0	30	2.5	22.3	73	9.5	846	0
Caribou, ME	23.7	37.6	-8	68	14.2	3.55	8.15	12.0	34.9	NA	2.9	21.0	79	11.1	1029	0
Casper, WY	21.8	44.3	-21	72	6.6	0.77	2.72	10.1	37.1	NA	7.2	14.1	53	14.4	957	0
Charleston, SC	47.2	69.5	15	88	6.9	2.49	10.56	T	T	63	12.6	11.1	52	8.1	233	35
Charleston, WV	36.3	57.3	6	88	12.0	3.59	9.12	2.4	25.8	NA	5.2	17.8	62	6.7	546	0
Charlotte, NC	41.5	62.6	11	85	7.5	3.23	8.17	0.1	2.5	61	11.9	11.6	52	7.2	391	0
Chicago, IL	31.6	48.4	-2	81	9.8	2.92	8.22	2.1	14.8	38	5.9	17.6	64	10.9	750	0
Cincinnati, OH	35.3	53.3	1	83	10.8	3.46	7.51	2.4	12.1	49	6.2	17.9	63	9.6	621	0
Cleveland, OH	35.0	50.0	0	82	11.0	3.17	8.80	5.2	23.4	31	3.1	21.0	66	11.9	672	0
Dallas, TX	45.4	66.8	17	89	5.7	2.29	9.91	0.2	5.0	62	12.3	11.7	56	10.6	275	11
Denver, CO	25.4	52.5	-18	80	5.2	0.87	3.21	8.1	42.6	65	10.4	10.1	44	8.4	780	0
Des Moines, IA	29.9	48.0	-10	82	6.9	1.79	7.10	2.7	14.0	49	7.4	15.6	66	11.5	780	0
Detroit, MI	32.2	48.1	0	81	11.2	2.67	5.68	3.5	11.8	35	4.2	19.0	65	10.9	744	0
El Paso, TX	38.4	66.4	1	87	2.7	0.44	2.50	1.0	12.7	82	17.3	6.3	33	8.1	378	0
Ely, NV	19.0	49.2	-15	75	5.1	0.67	1.82	5.2	17.3	66	9.8	12.1	46	10.0	924	
Grand Rapids, MI	30.2	45.8	-10	81	16.6	3.32	7.90	7.5	28.2	28	3.0	21.7	69	10.3	810	0
Hartford, CT	32.8	51.0	1	83	11.2	4.04	8.53	1.7	15.6	47	5.6	16.0	67	8.4	693	0
Helena, MT	20.7	42.4	-39	72	6.8	0.48	3.29	6.5	32.9	44	4.7	17.8	57	7.1	1002	0
Honolulu, HI	70.3	84.1	57	93	9.2	3.00	18.79	0.0	—	61	7.1	9.2	59	10.9	0	366
Houston, TX	49.6	72.4	19	89	8.6	3.79	14.10	T	T	53	10.2	12.0	59	7.7	181	61
Indianapolis, IN	34.1	51.9	-5	81	10.1	3.23	9.35	2.0	9.8	42	6.5	16.8	72	10.4	660	0
Jackson, MS	42.3	69.2	15	89	8.3	4.81	15.76	T	0.2	55	10.2	12.6	74	7.5	295	19
Jacksonville, FL	50.2	73.6	21	88	6.3	2.19	7.85	0.0	—	60	10.9	10.7	79	7.9	149	56
Kansas City, MO	33.6	52.6	1	83	7.3	1.92	9.52	1.0	9.1	50	8.8	14.8	62	11.3	657	0
Las Vegas, NV	42.6	67.4	15	87	2.1	0.43	2.22	0.1	4.0	80	15.7	6.9	27	7.5	304	0
Lexington, KY	37.0	54.9	-3	83	10.6	3.39	8.50	0.6	9.7	NA	7.2	16.2	62	10.3	570	0
Lincoln, NE	27.3	50.2	-15	85	5.9	1.27	7.14	2.7	12.6	54	8.2	14.6	66	10.2	786	0
Little Rock, AR	42.4	61.2	10	86	8.0	5.16	14.82	0.2	4.9	56	11.0	13.1	64	8.1	396	0
Los Angeles, CA	53.5	72.4	32	101	5.3	1.98	9.68	0.0	T	74	12.7	9.8	55	6.5	105	45
Lubbock, TX	36.5	63.2	-1	90	3.5	0.75	6.65	1.2	21.4	67	14.6	8.5	47	11.6	456	0
Marquette, MI	23.3	36.9	-13	74	15.4	2.89	8.25	15.7	42.0	28	NA	NA	NA	NA	1047	0
Medford, OR	35.5	52.6	10	80	12.3	3.23	8.62	0.5	11.4	NA	3.6	20.6	69	3.6	627	0
Memphis, TN	42.7	62.3	9	86	8.5	5.10	14.53	0.1	4.9	58	10.1	13.6	55	9.2	380	0
Miami, FL	66.7	80.4	36	91	8.2	2.66	17.72	0.0	—	66	8.1	8.1	61	9.5	6	264
Milwaukee, WI	30.7	44.7	-14	77	9.8	2.51	8.56	2.9	18.4	40	5.8	18.2	67	12.6	819	0
Minneapolis, MN	25.2	41.0	-17	77	8.2	1.55	5.29	6.6	46.9	39	5.5	17.7	65	10.9	954	0
Montgomery, AL	44.6	68.7	13	87	7.7	4.08	21.32	T	—	56	11.0	12.6	66	6.6	274	22
Nashville, TN	39.6	60.4	-1	85	9.5	4.12	9.04	0.5	9.2	49	8.7	14.6	60	8.4	450	0
New Orleans, LA	51.0	71.1	24	89	7.2	4.42	19.81	T	T	55	10.8	11.3	61	8.6	178	61
New York, NY	41.1	54.0	5	84	9.2	4.47	12.41	0.9	19.0	52	9.0	11.5	59	9.9	522	0
Norfolk, VA	43.8	61.2	9	86	8.0	2.85	9.67	T	3.5	58	10.7	11.4	69	10.6	380	0
Oklahoma City, OK	38.6	60.4	9	87	5.2	1.98	9.63	0.5	7.5	60	11.9	11.3	60	12.4	462	0
Orlando, FL	57.5	78.5	29	89	5.4	2.30	10.29	0.0	—	NA	10.7	9.1	55	8.6	54	144
Philadelphia, PA	37.6	55.1	8	81	9.3	3.34	9.06	0.6	13.4	52	7.3	13.6	55	9.6	558	0
Phoenix, AZ	46.9	74.3	25	96	2.5	0.54	3.61	0.0	0.1	83	17.5	6.3	28	5.4	159	27
Pittsburgh, PA	34.1	50.4	-1	82	13.0	2.85	4.70	3.8	32.3	38	4.0	19.7	62	9.8	681	0
Portland, OR	39.5	52.6	11	73	17.9	5.34	15.77	0.4	8.6	29	3.0	22.9	74	8.7	567	0
Providence, RI	34.9	53.0	6	82	11.0	4.43	11.01	0.6	10.2	50	8.4	14.9	69	10.5	630	0
Raleigh, NC	39.7	62.6	11	88	8.3	2.98	9.03	0.1	3.6	58	11.5	11.2	53	7.7	414	0
Reno, NV	26.7	53.8	1	80	5.0	0.87	4.65	2.3	14.0	71	9.3	12.8	42	5.5	741	0
Sacramento, CA	43.4	63.1	26	87	7.1	2.72	11.34	0.0	—	63	9.6	13.5	60	6.2	351	0
Salt Lake City, UT	30.9	50.8	-14	75	7.5	1.29	5.81	6.2	39.1	54	8.7	14.1	57	7.9	726	0
San Antonio, TX	48.8	71.9	21	94	6.3	2.62	9.46	0.0	0.3	55	10.8	12.2	55	8.9	180	42
San Diego, CA	53.9	69.9	36	97	4.7	1.45	5.82	0.0	—	74	14.6	7.6	57	5.8	109	19
San Francisco, CA	51.6	62.7	25	86	7.2	3.21	11.78	0.0	—	62	10.9	10.6	64	7.3	237	0
Seattle, WA	40.1	50.5	6	74	18.0	5.83	15.63	1.0	20.5	29	2.6	23.1	74	9.2	591	0
Sioux Falls, SD	22.6	43.4	-17	81	6.0	1.09	4.76	4.6	21.9	NA	7.1	15.6	70	11.5	960	0
Spokane, WA	28.8	41.4	-21	70	12.6	2.15	5.85	5.9	24.7	28	3.2	21.6	75	8.4	897	0
St. Louis, MO	37.7	54.7	1	85	9.4	3.28	9.95	1.5	12.7	46	8.8	14.5	69	10.0	564	0
Tampa, FL	57.2	77.7	23	90	5.4	1.77	6.12	0.0	—	65	11.8	9.0	56	8.5	70	148
Washington, DC	41.1	58.3	9	86	8.2	3.12	7.18	0.7	11.5	52	8.4	13.6	53	9.2	456	0
Wichita, KS	33.9	55.3	1	86	5.0	1.59	6.69	1.2	9.0	58	10.5	13.2	63	12.1	609	0

INTERNATIONAL DATA

	Avg High Temp	Avg Low Temp	Avg Precip	Days with Measurable Precip
Cape Town, South Africa	73	55	1.0	NA
Delhi, India	84	52	0.1	NA
Hong Kong, China	74	65	1.7	NA
Jerusalem, Israel	70	53	2.8	NA
London, United Kingdom	50	42	2.5	15
Manila, Philippines	87	71	4.5	12
Melbourne, Australia	71	51	2.3	12
Montreal, Canada	42	29	3.2	15
Moscow, Russia	33	26	1.4	13
Nairobi, Kenya	77	57	5.2	16
Paris, France	50	40	2.0	15
Rio de Janeiro, Brazil	79	68	4.1	13
Rome, Italy	61	49	5.1	11
Sydney, Australia	74	60	3.1	12
Tahiti, Society Islands	88	71	5.9	NA
Tokyo, Japan	60	43	3.8	8
Toronto, Canada	45	31	2.5	13

DECEMBER—U.S. CLIMATIC DATA

	TEMPERATURE (°F)					AMOUNT OF PRECIPITATION					MEAN			AVG	DEGREE DAYS (BASE 65°F)	
	AVERAGE		RECORD		MEAN DAYS OF	TOTAL		SNOWFALL		% OF POSSIBLE	CLEAR	CLOUDY	REL HUMID-	WIND SPEED		
	MIN	MAX	MIN	MAX	PRECIP	AVG	MAX	AVG	MAX	SUN	DAYS	DAYS	ITY	(MPH)	HEATING	COOLING
Albany, NY	18.2	34.9	-22	71	12.5	2.93	6.73	15.8	57.5	38	4.9	19.3	74	9.3	1194	0
Albuquerque, NM	23.1	47.5	-6	72	4.0	0.50	2.93	2.6	23.7	72	13.8	9.6	43	7.7	921	0
Amarillo, TX	23.7	50.1	-8	83	3.9	0.43	4.52	2.2	21.2	67	12.9	10.5	49	12.9	871	0
Anchorage, AK	10.0	22.5	-36	53	10.4	1.12	3.49	14.5	41.6	32	6.8	20.0	74	5.9	1510	0
Atlanta, GA	35.0	54.0	0	79	10.3	4.33	12.94	0.2	6.2	51	9.2	15.3	59	9.9	636	0
Atlantic City, NJ	26.3	45.3	-7	77	9.6	3.32	8.24	2.3	17.5	44	8.1	15.4	71	10.8	905	0
Bakersfield, CA	38.3	56.5	13	87	5.3	0.63	2.98	T	3.0	NA	7.2	16.4	62	5.0	543	0
Bismarck, ND	3.3	24.5	-43	66	7.8	0.51	1.71	6.8	21.7	47	6.7	17.4	69	9.5	1581	0
Boise, ID	22.5	37.7	-25	70	11.8	1.36	5.96	5.8	36.6	38	4.7	20.8	71	8.3	1082	0
Boston, MA	26.7	40.4	-17	76	11.7	4.01	9.74	7.8	27.9	52	8.8	14.9	60	13.5	973	0
Buffalo, NY	22.9	35.3	-10	74	19.7	3.67	8.71	22.6	82.7	27	1.2	23.6	74	13.4	1113	0
Burlington, VT	15.5	30.4	-29	67	15.0	2.42	5.95	19.4	56.7	32	2.8	22.4	72	9.8	1302	0
Caribou, ME	5.5	24.0	-31	58	14.7	3.22	7.97	23.8	59.9	NA	5.3	19.0	79	11.5	1556	0
Casper, WY	13.7	33.9	-41	61	7.5	0.66	3.71	11.2	62.8	NA	7.6	15.3	59	16.2	1277	0
Charleston, SC	40.7	61.6	8	83	8.5	3.15	10.56	0.1	3.8	59	9.5	14.3	55	8.6	439	11
Charleston, WV	28.0	46.0	-17	80	14.0	3.39	8.02	4.8	18.6	NA	4.6	19.8	66	7.2	868	0
Charlotte, NC	32.8	52.3	-5	79	9.7	3.48	11.24	0.6	15.0	58	10.2	14.9	57	7.4	694	0
Chicago, IL	19.1	34.0	-25	71	11.7	2.47	8.56	9.2	35.3	39	5.7	19.3	70	11.0	1190	0
Cincinnati, OH	25.3	41.5	-20	75	12.2	3.15	7.90	3.8	16.3	33	5.1	19.9	69	10.2	977	0
Cleveland, OH	24.5	37.4	-15	77	16.4	3.09	8.59	11.5	35.1	26	2.4	23.8	70	12.3	1057	0
Dallas, TX	36.3	57.5	-1	88	6.1	1.84	8.75	0.2	6.4	56	11.4	13.4	59	11.1	566	5
Denver, CO	17.4	44.5	-25	79	5.2	0.64	5.21	7.0	57.4	67	10.7	10.6	45	8.7	1054	0
Des Moines, IA	16.1	32.6	-22	69	7.9	1.32	3.72	6.7	26.9	45	6.9	17.6	71	11.4	1259	0
Detroit, MI	21.4	35.2	-24	69	13.7	2.82	6.00	10.5	34.9	30	3.4	21.2	71	11.3	1138	0
El Paso, TX	30.7	57.5	-5	80	3.6	0.57	3.94	1.3	25.9	78	15.3	8.3	36	8.1	648	0
Ely, NV	10.6	40.6	-29	67	6.3	0.70	2.11	7.3	22.3	65	8.9	14.2	55	10.0	1221	0
Grand Rapids, MI	20.7	33.5	-18	69	16.6	2.85	6.95	17.7	59.2	21	1.9	25.2	75	10.7	1175	0
Hartford, CT	21.3	37.5	-18	76	12.3	3.91	8.36	11.3	45.3	49	6.7	16.8	67	8.6	1101	0
Helena, MT	11.2	31.3	-40	64	8.1	0.59	4.64	8.7	46.4	41	4.0	20.2	64	6.9	1358	0
Honolulu, HI	67.0	81.2	54	89	10.2	3.80	17.29	0.0	—	58	8.2	9.6	61	10.7	0	282
Houston, TX	42.2	64.7	7	84	8.6	3.45	14.38	T	2.5	55	8.1	17.1	61	8.1	374	18
Indianapolis, IN	23.2	38.5	-23	74	11.8	3.34	7.72	4.9	27.5	39	5.1	19.7	76	10.5	1057	0
Jackson, MS	36.1	59.5	4	84	9.9	5.91	17.70	T	8.0	49	9.0	15.6	74	8.5	542	9
Jacksonville, FL	43.4	66.8	11	84	7.9	2.72	9.77	T	T	57	9.3	13.6	78	8.0	331	25
Kansas City, MO	21.9	38.8	-23	74	7.9	1.58	5.42	4.5	16.6	50	8.9	16.0	65	11.2	1073	0
Las Vegas, NV	33.9	57.5	11	78	2.5	0.38	1.78	0.1	2.0	77	14.6	9.6	33	7.2	598	0
Lexington, KY	27.6	44.2	-19	75	11.6	3.98	10.17	1.7	19.4	NA	6.0	19.0	68	10.9	902	0
Lincoln, NE	15.4	35.8	-27	75	5.5	0.88	4.03	5.6	22.3	52	7.9	15.8	71	10.3	1221	0
Little Rock, AR	33.7	50.7	-2	80	9.0	4.69	16.92	0.8	9.8	48	9.2	16.0	65	8.4	707	0
Los Angeles, CA	48.8	67.8	30	94	5.3	2.03	15.80	T	T	71	12.7	9.8	55	6.5	226	18
Lubbock, TX	27.2	54.1	-2	85	3.9	0.53	2.70	1.8	9.9	65	13.5	10.5	48	11.9	756	0
Marquette, MI	9.9	24.6	-28	60	18.4	2.61	8.65	26.7	89.5	27	NA	NA	NA	NA	1479	0
Medford, OR	31.2	44.3	-10	72	14.5	3.32	12.72	1.4	12.2	NA	1.9	25.1	76	3.6	843	0
Memphis, TN	34.8	52.5	-13	81	9.9	5.74	13.81	0.8	14.3	50	8.8	16.3	62	9.9	660	0
Miami, FL	61.5	76.7	30	87	6.5	1.83	9.03	0.0	—	67	9.4	8.8	59	9.2	41	168
Milwaukee, WI	17.5	31.2	-22	68	11.2	2.33	5.42	10.6	49.5	37	6.0	19.1	72	12.4	1259	0
Minneapolis, MN	10.2	25.5	-29	68	9.1	1.08	4.27	9.6	33.2	40	6.1	18.4	70	10.4	1460	0
Montgomery, AL	38.7	60.2	5	85	10.1	5.20	11.35	T	11.0	51	9.5	15.5	66	7.2	492	11
Nashville, TN	30.9	50.2	-10	79	10.8	4.61	13.53	1.7	13.2	42	7.2	17.0	63	8.9	760	0
New Orleans, LA	44.8	64.3	11	84	10.1	5.75	14.43	0.1	2.7	53	8.1	15.3	66	9.1	349	23
New York, NY	30.7	42.5	-13	75	10.4	3.91	9.98	5.6	29.6	49	8.9	13.0	60	10.4	880	0
Norfolk, VA	35.4	52.2	-5	80	9.1	3.23	8.72	1.0	14.7	57	9.4	14.4	68	11.0	657	0
Oklahoma City, OK	28.6	49.9	-8	86	5.3	1.40	8.14	1.6	9.0	59	11.7	13.2	61	12.6	797	0
Orlando, FL	51.3	72.9	20	90	6.1	2.15	12.63	0.0	—	NA	9.8	12.0	57	8.6	164	74
Philadelphia, PA	28.1	43.4	1	73	10.2	3.38	8.47	3.8	22.4	49	7.5	15.1	59	10.0	905	0
Phoenix, AZ	40.2	66.4	22	88	3.8	0.83	3.98	T	0.1	77	15.1	9.7	34	5.2	368	6
Pittsburgh, PA	24.4	38.6	-12	74	16.4	2.92	8.51	8.3	36.3	29	2.6	22.6	67	10.4	1039	0
Portland, OR	34.8	45.6	3	65	19.3	6.13	20.14	1.4	34.1	21	1.8	26.2	79	9.6	769	0
Providence, RI	24.4	41.2	-12	77	12.3	4.38	10.75	7.1	26.7	52	8.3	14.8	67	10.9	998	0
Raleigh, NC	32.4	52.7	4	80	9.0	3.24	7.78	0.8	10.6	55	10.3	13.8	55	8.0	694	0
Reno, NV	19.9	45.5	-16	70	6.0	0.99	5.25	4.3	33.8	64	8.2	15.0	52	5.2	1001	0
Sacramento, CA	37.8	52.7	17	72	9.2	2.51	13.40	T	T	46	7.6	17.6	71	6.8	611	0
Salt Lake City, UT	21.6	37.8	-21	68	9.4	1.40	4.37	12.3	35.2	43	6.3	18.1	70	7.5	1094	0
San Antonio, TX	40.8	63.5	6	90	7.4	1.51	13.96	0.0	3.0	50	9.7	15.1	57	8.7	409	12
San Diego, CA	48.8	66.1	32	88	5.7	1.57	9.26	T	T	72	13.7	9.6	55	5.5	243	7
San Francisco, CA	47.0	56.4	20	76	9.9	3.10	15.16	0.0	3.5	53	9.2	14.1	68	6.9	412	0
Seattle, WA	35.8	45.1	6	65	20.0	5.91	15.33	2.8	22.1	20	2.0	25.5	77	9.8	760	0
Sioux Falls, SD	8.6	28.0	-31	63	6.3	0.70	2.97	7.6	41.1	NA	7.3	16.7	75	10.7	1448	0
Spokane, WA	21.7	33.8	-25	60	15.4	2.42	5.13	15.6	42.7	21	2.8	24.4	82	8.6	1153	0
St. Louis, MO	26.0	41.7	-16	76	9.4	3.03	10.90	3.6	26.3	42	7.1	17.2	74	10.4	964	0
Tampa, FL	52.3	72.1	18	86	6.4	2.15	15.57	0.0	T	62	10.2	11.4	59	8.6	171	84
Washington, DC	31.7	47.0	-13	79	9.1	3.12	7.56	3.3	24.2	47	8.4	16.1	56	9.5	794	0
Wichita, KS	23.0	43.0	-16	83	5.8	1.20	4.71	3.3	16.3	58	9.9	14.1	66	12.21	992	0

INTERNATIONAL DATA

	AVG HIGH TEMP	AVG LOW TEMP	AVG PRECIP	DAYS WITH MEASURABLE PRECIP
Cape Town, South Africa	76	58	0.8	NA
Delhi, India	73	46	0.4	NA
Hong Kong, China	68	59	1.2	NA
Jerusalem, Israel	59	45	3.4	NA
London, United Kingdom	45	38	1.9	15
Manila, Philippines	86	69	2.1	8
Melbourne, Australia	75	55	2.3	10
Montreal, Canada	27	13	3.4	18
Moscow, Russia	25	17	1.4	16
Nairobi, Kenya	78	56	3.0	11
Paris, France	44	36	2.0	16
Rio de Janeiro, Brazil	82	71	5.4	14
Rome, Italy	55	44	3.7	10
Sydney, Australia	77	63	3.1	13
Tahiti, Society Islands	88	72	9.8	NA
Tokyo, Japan	52	33	2.2	5
Toronto, Canada	33	19	2.6	15

CONTRIBUTORS

Dr. D. James Baker is an oceanographer and science and management consultant working with the Intergovernmental Oceanographic Commission of UNESCO in Paris and the H. John Heinz Center for Science, Economics and the Environment in Washington, D.C. He was the Administrator of the U. S. National Oceanic and Atmospheric Administration (NOAA) from 1993 to 2001. He is chair of the review committee for the World Climate Research Program and is a member of the International Science Steering Committee for the Census of Marine Life. He previously served as President of the Academy of Natural Sciences in Philadelphia and of Joint Oceanographic Institutions Incorporated in Washington, D.C.; as Dean of the College of Ocean and Fishery Sciences at the University of Washington; and as an associate professor at Harvard University. He is the author of the book *Planet Earth: The View from Space*, and has written extensively about climate, oceanography, and space technology issues.

Marlene Bradford is an independent scholar and weather historian living in Bryan, Texas. She is the author of *Notable Natural Disasters,* published by Salem Press, and *Scanning the Skies: A History of Tornado Forecasting,* published by University of Oklahoma Press.

Jay Brausch is a writer, photographer and longtime observer and recorder of meteorological and astronomical events from western North Dakota. He has documented elusive weather phenomena such as sunspots, auroras, and noctilucent clouds in drawings and photography since 1993. His photographs of these rare events have won numerous science and photographic awards and have been published in noted publications throughout the world.

Kathleen Cain is the author of *The Cottonwood Tree: An American Champion,* recently published by Johnson Books. She has also written and published poetry, essays, and feature articles in both literary and popular magazines. She has been a contributing editor for *The Bloomsbury Review* since 1982.

Randy Cerveny is a President's Professor of Geographical Sciences at Arizona State University and a contributing editor of the magazine *Weatherwise*. He is also the author of the book *Freaks of the Storm* published by Thunder's Mouth Press.

Nick D'Alto is a writer whose news and feature stories on weather, science, and technology have appeared in magazines, including *Weatherwise, Air & Space,* and *American Heritage.*

Ed Darack is an independent writer and photographer who covers a wide range of topics, but has a particular interest in all subjects weather. A contributing editor to *Weatherwise* magazine, he has been publishing weather-related articles for over a decade. His images are represented through Getty Images and Science Faction Images.

John A. Day is a retired physics and meteorology professor at Linfield College, Oregon, and the author of more than seven books on meteorology, including *The Book of Clouds.* His initial interest in clouds was technical, as he learned to forecast their appearance and development as they impacted Pan Am Clipper flights. He later began photographing clouds and has an extensive collection of cloud photography. He and his wife Mary of 70 years live in McMinnville, Oregon.

Warren Faidley is journalist, photographer, cinematographer, and adventurer who specializes in the coverage of severe weather, global warming, and natural disasters. Over the past 20 years, he has experienced and survived some of the earth's most breathtaking and sometimes violent weather, including both an F5 tornado and a Category 5 hurricane. He often serves as an on-air consultant for the media, including Fox News, CNN, and others. Warren is the CEO and founder of Weatherstock Inc., a stock agency specializing in severe weather footage and photographs.

Scott Hadly is a journalist whose work has appeared in several publications, including the *Los Angeles Times,* the *San Francisco Examiner,* the *Arizona Republic,* and the magazine *California Journal.* Hadly has won awards for his work in each of the 15 years he's been a reporter. He's documented the decline of ranching in southern California, corporate influence on academic research with the University of California system, local political corruption, and environmental crimes. A lifelong surfer and outdoorsman, Hadly now works as a reporter for the *Ventura County Star.* He lives in Santa Barbara with his wife and two children.

Kate James is a writer and photographer living in the Colorado Rockies. Her most recent photo essays are of the Pawnee Buttes of Colorado and a small Malaysian village called Serian. Her most recent written work appeared in *Western Horseman* and her photographs have appeared in *Shots.*

Tim Kern is a writer and consultant who has contributed to dozens of periodicals, including *SUCCESS* magazine and *The Farmer's Almanac.*

Jim LaDue is a meteorologist with the National Weather Service. He trains other meteorologists on how to use radar and satellite data so that they may produce more accurate forecasts and severe weather warnings. He also develops new techniques for combining satellite and radar data. His hobbies include canoeing, skiing, hiking, storm chasing, and photography.

Kim Long is a Denver-based writer and the former editor of the *Weather Calendar* from Accord Publishing. He is also the author of dozens of nonfiction books on topics ranging from business and consumer trends to tree squirrels and the moon. His latest book is *The Almanac of Political Corruption, Scandals, and Dirty Politics,* published by Bantam Doubleday Dell.

Walter A. Lyons, Ph.D., is a Certified Consulting Meteorologist with 40 years' experience in research for government agencies, operational forecasting, broadcasting, and forensic investigations. His current interests are in producing and providing weather-related DVDs for educators and weather enthusiasts. He is a past president of the American Meteorological Society and president elect of the National Council of Industrial Meteorologists.

Gregory McNamee is a writer and frequent editor of and contributor to the *Weather Calendar*. He writes on science, geography, and culture for many publications, including *Encyclopedia Britannica*. His most recent book is *Moveable Feasts: The History, Science, and Lore of Food.*

Vince Miller is an Oklahoma City-based meteorologist who has contributed the daily weather trivia and annual climatic data for the *Weather Guide Calendar* for the past 20 years. He worked for the Weather Channel for 10 years and was the chief meteorological advisor to the film "Twister." He is currently an adjunct meteorology professor at Southeastern Oklahoma State University.

Rich Reid is a photographer and writer specializing in commercial and editorial assignments. As an instructor at Brooks Institute of Photography, he travels with students to photograph landscape and weather events and conducts photographic tours in Alaska. His work is represented through the National Geographic Image Collection.

Fred Schaaf is the author of 13 books on astronomy. His most recent books are *The 50 Best Sights in Astronomy and How to See Them* and *The Brightest Stars,* both published by John Wiley & Sons. Schaaf has written two ongoing columns in *Sky & Telescope* magazine for more than 15 years and teaches astronomy at Rowan University in Glassboro, New Jersey.

Larry Sessions, a former planetarium director, teaches and writes about astronomy, and is a contributing writer for *SPACE.com* and *Earth & Sky* magazine. He also writes a blog for *Earth & Sky* on "Clouds and Cosmos: The Atmosphere and Beyond!"

Gregory J. Stumpf is a meteorologist at the National Weather Center and lives in Norman, Oklahoma. He helps develop and transfer technology to the National Oceanic and Atmospheric Administration's (NOAA) National Weather Service (NWS) severe weather warning operations, including Doppler radar algorithms. He also has made a 23-year career of chasing storms and tornadoes on the Great Plains, having witnessed over 150 tornadoes.

Randy Welch has been a Denver-based freelance writer for more than 20 years, covering a wide variety of styles and topics. He also works part-time for the Federal Emergency Management Agency (FEMA), traveling to disasters to work with the media—which keeps him in tune with wild weather events of all kinds.

PHOTOGRAPHY CREDITS

All photographs are used by permission of the artists and/or organizations and are copyrighted in the names as listed below. They may not be reproduced in any form without the permission of the artist or agency.

Cover

A. T. Willett/lightningsmiths.com

Front Matter

P. 1: Gene Rhoden/Weatherpix Stock Images; pp. 2–3: Warren Faidley/Weatherstock®.com; pp. 4–5: Rod Hanna; pp. 6–7: Chuck Conway; pp. 8–9: Rod Hanna; p. 10: Tsuyoshi Nishiinoue; p. 11: E. R. Degginger/Dembinsky Photo Associates; p. 12: A. T. Willett/lightningsmiths.com; pp.12–13: Gene Rhoden/Weatherpix Stock Images; pp. 14–15: Fred Hirschmann; p. 17: NASA; p. 23: F. Jocelyn Augustino/Weatherstock®.com

Forecasting

P. 24, top left: Kent Wood/Peter Arnold, Inc.; p. 24, top right: NASA; p. 24, bottom left: Colin McNulty/Small World Images; p. 24, bottom right: E. R. Degginger/Dembinsky Photo Associates; p. 25: Adam Block; pp. 26: Jack Dykinga; pp. 28–29: Jim Reed/Getty Images; p. 29: Courtesy of NOAA; p. 30: Lon Curtis; p. 31: A. T. Willett/lightningsmiths.com; p. 32: Dick Dietrich/Dietrich Leis Stock Photography; p. 35: Charles A. Doswell III; p. 36: Eugene McCaul Jr.; p. 39: Jason Lafontaine/Painet, Inc.; p. 40: Galen Rowell/Mountain Light; p. 41: British Crown © 2007, the Met Office; p. 43: Weatherstock®.com; pp. 44–45, 47: Ed Darack; p. 48: Rich Reid; p. 51–53: Warren Faidley/Weatherstock®.com

Light

P. 54, top left: Jim Reed/Getty Images; p. 54, top right: Fred Hirschmann; p. 54, bottom left: Steven J. Mueller; p. 54, bottom right: NASA; p. 55: NASA; p. 56: Charles A. Doswell III; pp. 58–59: Rod Hanna; p. 60: R. Hadian/U.S. Geographical Survey; p. 62: Richard Kaylin/Getty Images; p. 63: Hank Baker/weatherpic.com; p. 64: Dan Heller/danheller.com; p. 65: Jack Dykinga; p. 66: Gene Rhoden/Weatherpix Stock Images; p. 67, top row: Heather Gravning; p. 67, bottom row: David Ewolt; p. 68: Thomas Wiewandt/Getty Images; p. 71: Dan Heller/danheller.com; p. 72: Douglas Keister; p. 73: Keith Kent/Peter Arnold, Inc.; p. 74: Fred Hirschmann; pp. 76–77: Jay Brausch; p. 78: Galen Rowell/Mountain Light; p. 79: NASA; p. 80: Sigurdur Stefnisson; p. 81: Fred Hirschmann; pp. 82–83: Steve Bloom/stevebloom.com; p. 84, 86: Kenneth D. Langford; p. 87: Paul Neiman; pp. 88–89: E. R. Degginger/Dembinsky Photo Associates

Wind

P. 90, top left: Brian A. Morganti/stormeffects.com; p. 90, top right: E. R. Degginger/Dembinsky Photo Associates; p. 90, bottom left: Warren Faidley/Weatherstock®.com; p. 90, bottom right: Eugene McCaul Jr.; p. 91: Gene Rhoden; p. 92: Hank Baker/weatherpic.com; p. 94: Jim Reed/Getty Images; p. 95: Eric Nguyen; p. 96: Karen Leszke & Gene Rhoden/Weatherpix Stock Images; p. 99: Terry Mosher; pp. 100–101: Gene Rhoden; p. 102: Vic Bider/Getty Images; p. 104: Karen Leszke & Gene Rhoden/Weatherpix Stock Images; p. 105: Eric Nguyen; p. 106: Gordon McNorton; p. 108: Bob Firth/firthphotobank.com; p. 109: Paul R. Donovan; p. 110: Jahi Chikwendiu/*The Washington Post*; p. 113: David Pabst; p. 114: Orbimage; p. 115: Stan Osolinski/Dembinsky Photo Associates; p. 116: E. R. Degginger/Dembinsky Photo Associates; p. 119: Gary Kaufman; p. 120: Gary Felton/garyfelton.com; p. 121: Ken Araujo; p. 122: Warren Faidley/Weatherstock®.com; p. 123: Courtesy of NOAA; p. 125: Warren Faidley/Weatherstock®.com; p. 126: Kyle Niemi, courtesy of U.S. Coast Guard; p. 127, 129: Warren Faidley/Weatherstock®.com

Water

P. 130, top left: Chuck Conway; p. 130, top right: Jim Mastro; p. 130, bottom left: Rich Reid; p. 130, bottom right: Galen Rowell/Mountain Light; p. 131: Colin McNulty/Small World Images; p. 132: Curtis Martin; pp. 134–135: Susan Strom; p. 136: Howie Garber/Wanderlust Images; p. 137: Carol Simowitz; p. 139: Tom Brownold; pp. 140–141: Alan Moller; p. 142: Carter E. Gowl; p. 145: Hank Baker/weatherpic.com; p. 146: David L. Sladky; p. 147: Copyright © 2007 University Corporation for Atmospheric Research; p. 149: Alan Moller; pp. 150–151: Viveca Venegas/imagenesdemiscaminos.com; p. 152: Barry Slade; p. 156: Charles A. Doswell III; p. 157, top left: Jeff Foott/Getty Images; p. 157, top middle: Jorn Olsen; p. 157, top right: Warren Faidley/Weatherstock®.com; p. 157, bottom left: Brooks Martner; p. 157, bottom middle: Richard Kaylin/Getty Images; p. 157, bottom right: Stan Osolinski/Dembinsky Photo Associates; p. 158: Kathleen Norris Cook; p. 159: Gregory Thompson; p. 160: Galen Rowell/Mountain Light; p. 161: Fred Hirschmann; p. 162: Tom Vezo/tomvezo.com; p. 163: Paul Neiman; p. 164: Galen Rowell/Mountain Light; p. 166: Fred Hirschmann; p. 167: Ed Darack; p. 168: Layne Kennedy; p. 169: Stephen Graham/Dembinsky Photo Associates; p. 170: Rich Reid; p. 171: *Bentley's Snowflakes CD-Rom & Book*, Dover Publications, Inc.; p. 172: Jonathan Chester/extremeimages.com; p. 173: Tom Bean; p. 175: Jonathan Chester/extremeimages.com; p. 176: Joan Myers; p. 177: Jim Mastro; p. 178: Colin McNulty/Small World Images; p. 179: Norbert Wu; pp. 180–181: Jonathan Chester/extremeimages.com

Climate Change

P. 182, top left: Jonathan Chester/extremeimages.com; p. 182, top right: Astromujoff/Getty Images; p. 182, bottom left: Dean Conger/Getty Images; p. 182, bottom right: Ashley Cooper/Corbis; p. 183: Liu Liqun/Corbis; p. 184: George Steinmetz/Corbis; p. 186: Charles & Josette Lenars/Corbis; p. 187: Ashley Cooper/Corbis; p. 188: Dean Conger/Getty Images; p. 189, top left: Bob Firth/firthphotobank.com; p. 189, top right: Ed Darack; p. 189, bottom left: Fred Hirschmann; p. 189, bottom right: Rich Reid; p. 191: Michael Hanschke/dpa/Corbis; p. 192, top: Altaf Qadri/epa/Corbis; p. 192, bottom: Ashley Cooper/Corbis; p. 193: James Randklev/Corbis; p. 194: Jonathan Chester/extremeimages.com; p. 195: Michael Fiala/zefa/Corbis; p. 197: NASA; p. 198, top: Kennan Ward/Corbis; p. 197, bottom: Kazuyoshi Nomachi/Corbis; p. 198, 199: Jonathan Chester/extremeimages.com; p. 200: Remi Benali/Corbis; p. 201: Diego Azubel/epa/Corbis; p. 202: Yves Gellie/Corbis; p. 203: Michael Reynolds/epa/Corbis; p. 204, top: Remi Benali/Corbis; p. 204, bottom: Nic Bothma/epa/Corbis; p. 205: Bruno Fert/Corbis; p. 206: Liu Liqun/Corbis; p. 207: Michael Reynolds/epa/Corbis; p. 208, top: Barry Lewis/Corbis; p. 208, bottom: Justin Guariglia/Getty Images; p. 211: Barry Lewis/Corbis; pp. 212–223: Jonathan Chester/extremeimages.com; p. 224, top left: Brian A. Morganti/stormeffects.com; p. 224, top right and bottom left: A. T. Willett/lightningsmiths.com; p. 224 bottom right: Tom Bean

INDEX

Italics indicate pages with photos.

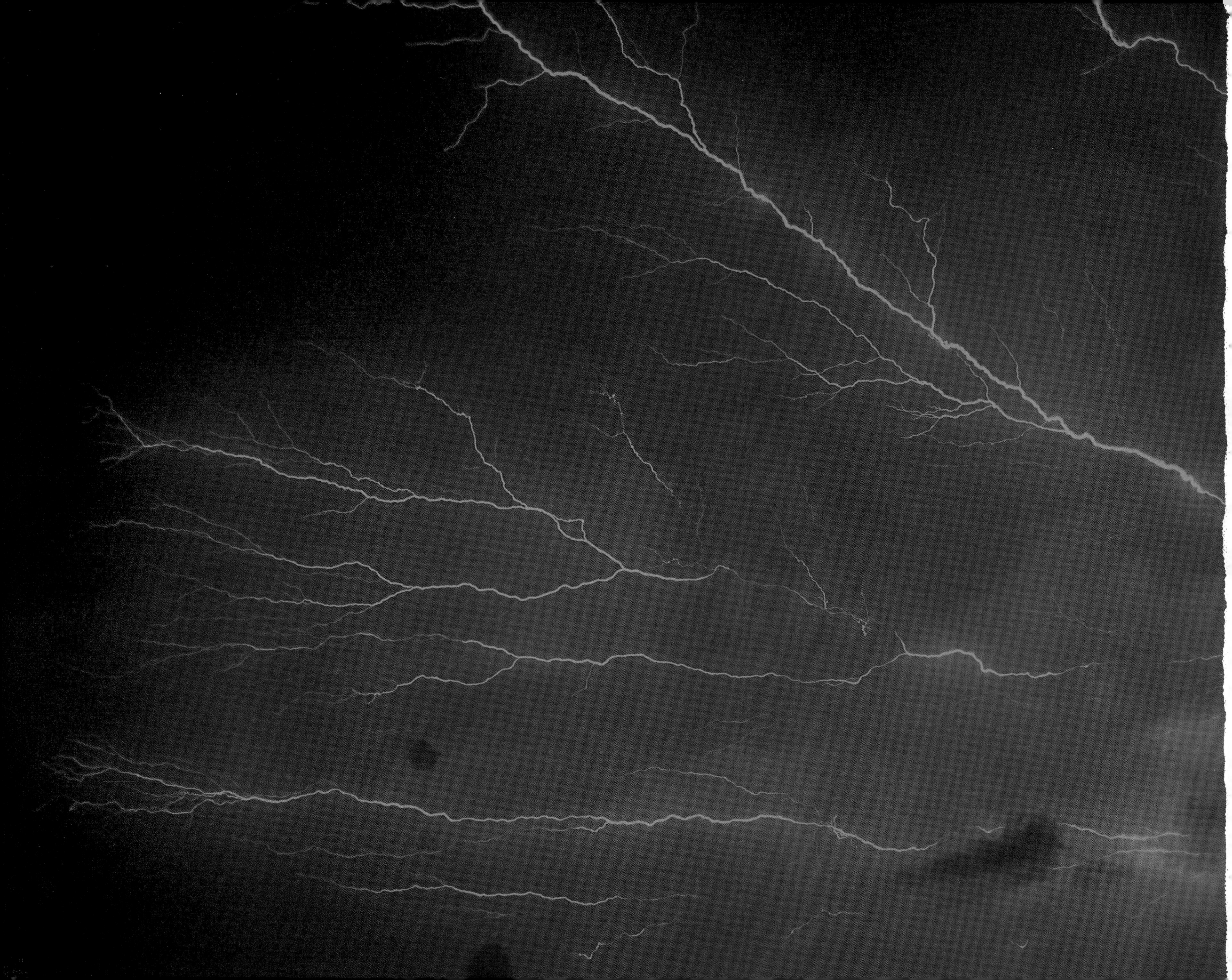